How to Measure

Training Results

How to Measure Training Results

A Practical Guide to Tracking
the Six Key Indicators

JACK J. PHILLIPS
RON DREW STONE

McGraw-Hill
New York   San Francisco   Washington, D.C.
Auckland   Bogota   Caracas   Lisbon   London
Madrid Mexico City   Milan   Montreal   New Delhi
San Juan   Singapore Sydney   Tokyo   Toronto

McGraw-Hill

A Division of The **McGraw·Hill** Companies

9   0   FGR/FGR   0   6   5

ISBN 0-07-138792-7

Printed and bound by Quebecor World / Fairfield.

McGraw-Hill books are available at special quantity discounts to use as premiums and sales promotions, or for use in corporate training programs. For more information, please write to the Director of Special Sales, Professional Publishing, McGraw-Hill, Two Penn Plaza, New York, NY 10121-2298. Or contact your local bookstore.

 This book is printed on acid-free paper.

Contents

♦ ♦ ♦

Acknowledgments

I want to thank all the facilitators who have used the materials in this book to conduct a two-day workshop on measuring return on investment. Since 1993, over 500 workshops have been conducted with 10,000 participants. As expected, the facilitators provided helpful input to improve both the materials and the workshops. A special thanks goes to Patti, my partner and spouse, who always adds value to our programs.

—Jack Phillips

I want to thank several of my professional associates who have dedicated their careers to training and who have taken a special interest in evaluation. In addition to my coauthor Jack Phillips, they are: Doug Hoadley, Chuck Bonovitch, Gary Parker, Jim Wright, Patrick Whalen, and Bruce Nichols. A special thanks to my wonderful spouse, Jo Ann, who has persistently encouraged me to complete this book; my daughter, Ronda; my granddaughters, Stacy, Mandy, and Amber; and my great granddaughter, Madison.

—Ron Stone

♦ ♦ ♦

Introduction

THE INCREASED DEMAND FOR PROVEN RESULTS

Increasingly, organizational stakeholders who fund training are interested in demonstrated results—measures of how training expenditures contribute to the organization. This interest stems from two things. In the private sector, global competition and investors are increasing the demand for accountability for results, as well as prudent expenditures. Government sectors are seeing the same pressure to contain budgets, expend funds wisely, and achieve greater results from expenditures. In addition, federal organizations must practice fiscal management and accountability in alignment with the requirements of the Government Performance and Results Act (GPRA).

HOW THIS BOOK CAN HELP YOU

This book is written for training practitioners and for anyone who is interested in using practical evaluation techniques to assess, improve, and report on training programs and results. This book provides techniques, tools, worksheets, and examples that you can use to follow a systematic process to plan and conduct credible evaluations of your training programs. It serves as a guide in planning and implementing evaluations at five different levels of results. The first two levels are of primary interest to the stakeholders involved in developing and presenting training and development programs. The third level is of primary interest to learner-participants and their immediate managers. The fourth and fifth levels and, to some extent, the third level are of primary interest to the executives and stakeholders who fund the training. This five-level framework for evaluation is described in detail throughout this book.

This book also provides you with ten standards for collecting and analyzing data. Overall, this book is a resource for understanding the evaluation methodology and tools needed to successfully implement the ROI Process in your organization. Thorough review of and use of the principles and methodology presented should allow you to do the following:

- Develop objectives for and develop and implement an evaluation plan for a specific training program
- Select appropriate data-collection methods for assessing impact
- Utilize appropriate methods to isolate the effects of the training
- Utilize appropriate methods to convert hard and soft data to monetary values
- Identify the costs of a training program
- Analyze data using credible methods
- Calculate the return on investment
- Make cost-effective decisions at each of the five evaluation levels
- Use data-based feedback to improve the effectiveness of training programs and discontinue ineffective programs
- Collect and report the type of performance data that will get the attention of senior management
- Present the six types of data that are developed from the ROI Process
- Convince stakeholders that your program is linked to business performance measures

- Link and enhance the implementation of training to improve organizational results
- Improve the satisfaction of your stakeholders

Organizations also may use this book to facilitate small-group discussions in order to help prepare internal staff members to conduct program evaluations.

WHAT THIS BOOK CONTAINS

This practical guide will help you implement a process to measure the results (up to and including return on investment) of training and performance-improvement programs in your organization. It allows you to take advantage of 20 years of experience in the application of a proven, systematic evaluation process. For ease of use, this book contains the following:

- A *model of the evaluation process* to guide you in understanding each of the steps involved (e.g., planning, data collection, assessment, financial calculations, and reporting).
- *Text* to provide you with the fundamentals of measurement and evaluation in the systematic ROI Process. Concepts and practical approaches are provided to set the stage for the planning and use of the evaluation tools.
- *Case illustrations* that allow you to see some of the concepts and practices in action in organizational settings. As extensions of the core text in the chapters, these cases are used sparingly and present situations that challenge your thinking about how the concepts should be applied. Additional commentary clarifies thought-provoking issues.
- *Worksheets* and job-aids that will help you to apply the principal tools for each of the components of the evaluation process.
- *Examples* of completed worksheets for those that are not self-explanatory. These examples illustrate how the worksheets are actually used.

- *Checklists* to help you address pertinent questions and issues related to the application of each key component of the measurement process.
- *Figures* to highlight, illustrate, and categorize learning points. Many of these are examples that clarify or add to the text.

WHY MOST TRAINING ISN'T MEASURED

Historically, training has been measured from the perspective of what transpired during the training program: did the participants enjoy the experience; was the content relevant; did learning occur? This approach is still used in many organizations, and a great deal of training is not measured beyond participant-reaction smile sheets and self-reported learning, which are easy to complete and tend to reflect positive results. Putting it simply, the training function and the participants often have not been held accountable for the transfer of learning to the work setting and the impact on key organizational measures.

Too often, training has been viewed as either a line-management responsibility or a responsibility of the HR or training department. The truth is that it is joint responsibility.

Senior management stakeholders have not asked enough questions about results. This may be because training costs are budgeted and allocated in ways that create indifference from line management and others. It may be because management has bigger fish to fry or because the training staff feels that training participants, line managers, and others will not cooperate in providing the data necessary to measure results. Often the training staff, managers, and others have been led to believe that the effects of training cannot be measured credibly, i.e., that they cannot be isolated from the influence of other performance-improvement factors or that it is too difficult or too resource intensive to measure the effects of training. Depending on the organization and the culture, one or more of these factors contribute to the lack of evidence that training brings benefits to the organization that are greater than the costs incurred.

CREDIBLE MEASUREMENT IS POSSIBLE

This book shows that the organizational impact of training can be measured with credibility and reasonable allocation of resources. For 20 years, Jack Phillips, the coauthor of this book, has been using the ROI Process to demonstrate the feasibility and methodology of measuring the influence of training on organizations. The other coauthor, Ron Stone, joined Jack Phillips in 1995 and has made many contributions to perfecting the ROI Process.

The ROI Process has proven to be a flexible and systematic methodology that others can learn and implement to measure and assess the impact of training programs. The process relies on a practical evaluation framework that can be applied to yield consistent and credible study results. The methodology, worksheets, and other tools provided in this book greatly simplify the decisions and activities necessary to plan and implement such a study. The practical experience of the authors in using the process in the private and government sectors provides many insights for putting the evaluation process into practice.

There are many choices to be made when deciding how to collect data, analyze data, isolate the effects of training, capture costs, convert data to monetary values, identify intangibles, and calculate the return on investment. This is good news because it means that you can learn to use this process and feel comfortable that the methodology will withstand the scrutiny of the stakeholders in your organization. It is the methodology you use that others will question when they view or hear about your evaluation results. You must learn the methodology and never compromise it. It is your credential to successful measurement. The methodology guides you in making the right choices. As you will learn and come to trust, these choices are almost always situational, and the process has the inherent flexibility to account for a wide variety of differences. This flexibility may be the most powerful aspect of the process, and it is one of the most appealing factors to the many practitioners worldwide that have used it. So read, plan, apply the process, and learn from your experiences.

How to Measure Training Results

1

The Need for and Benefits of Measurement and Evaluation of Training Outcomes

WHY MEASUREMENT AND EVALUATION ARE NECESSARY

Much has been written about the need for training and performance-improvement professionals to become more accountable and to measure their contributions. The organization funds training at the expense of other organizational needs, and the results influenced by training can be elusive without a focused evaluation effort to address the outcomes. Just as learning initiatives must include the various stakeholders, so too must the evaluation effort include the stakeholders of the organization. In essence, the training function must be a business partner in the organization in order to successfully deliver its products. Most observers of the field have indicated that for performance practitioners to become true business partners, three things must be in place.

1. Training and performance-improvement initiatives must be integrated into the overall strategic and operational framework of the organization. They cannot be isolated, event-based activities, unrelated to the mainstream functions of the business.

2. There must be a comprehensive measurement and evaluation process to capture the contributions of human resource development and establish accountability. The process must be comprehensive, yet practical, and feasible as a routine function in the organization.

3. Partnership relationships must be established with key operating managers. These key clients are crucial to the overall success of the training function.

Most training executives believe their function is now an important part of the business strategy. During the 1990s, and continuing into the twenty-first century, training and performance improvement have become a mainstream function in many organizations. The training executives of these organizations emphasize the importance of successfully establishing partnerships with key management and report that tremendous strides have been made in working with managers to build the relationships that are necessary. They report fair progress in the achievement of integrating training into the overall strategic and operational framework of the organization. However, they indicate that there has not been progress on the second condition: a comprehensive measurement and evaluation process—at least not to the extent needed in most organizations. This book is devoted to presenting the principles and tools necessary to allow practitioners to implement a comprehensive measurement and evaluation process to improve results in their organization. The installation of a comprehensive measurement and evaluation process will quite naturally address the other two items as well. The comprehensive measurement of training will provide for a closer link to the organization's strategic goals and initiatives. Measurement will also allow line managers to see the results as well as the potential from training efforts, and this will lend itself to stronger partnerships.

Measurement will continue to be necessary as long as the drivers for accountability exist. Some of the current drivers for accountability are operating managers' concern with bottom line, competition for funds and resources, accountability trend with all functions, top-management interest in ROI, and continuing increases in program costs. In the final analysis, the real issues behind accountability are the external forces of competition. In the business sector it is the competitive nature of the world marketplace. In government and nonprofit organizations, it is the competition for funds and resources to achieve the primary mission.

A FRAMEWORK FOR EVALUATION WITH SIX TYPES OF MEASURES

Measurement and evaluation are useful tools to help internalize the results-based culture and to track progress. When looking for evidence of accountability in training, the question of what to measure and what data to review is at the heart of the issue.

Applying the framework presented in this chapter, along with the ROI (return on investment) process, involves five types of data (associated with five levels of measurement) and a sixth type of data represented by intangible benefits. These can be used to measure training and educational programs, performance-improvement programs, organizational change initiatives, human resource programs, technology initiatives, and organization development initiatives. (For consistency and brevity, we use the term "training programs" throughout most of this book.)

The fifth level in this framework is added to the four levels of evaluation developed for the training profession almost 40 years ago by Donald Kirkpatrick.[1] The concept of different levels of evaluation is helpful in understanding how the return on investment is calculated.

Table 1.1 shows the modified version of the five-level framework as well as the intangible dimension.

Table 1.1. Five levels and six types of measures.

EVALUATION FRAMEWORK		
LEVEL AND TYPE OF DATA	FOCUS OF THE DATA	SUMMARY OF HOW THE DATA IS USEFUL
Level 1: Reaction and/or satisfaction and planned action	Focus is on the training program, the facilitator,and how application might occur	Reaction data reveals what the target population thinks of the program—the participants' reactions to and/or satisfaction with the training program and the trainer(s). It may also measure another dimension: the participants' planned actions as a result of the training, i.e., how the participants will implement a new requirement, program, or process, or how they will use their new capabilities. Reaction data should be used to adjust or refine the training content, design, or delivery. The process of developing planned actions enhances the transfer of the training to the work setting. Planned-action data also can be used to determine the focal point for follow-up evaluations and to compare actual results to planned results. These findings also may lead to program improvements.
Level 2: Learning	Focus is on the participant and various support mechanisms for learning	The evaluation of learning is concerned with measuring the extent to which desired attitudes, principles, knowledge, facts, processes, procedures, techniques, or skills that are

Table 1.1. Five levels and six types of measures (*continued*).

EVALUATION FRAMEWORK		
LEVEL AND TYPE OF DATA	**FOCUS OF THE DATA**	**SUMMARY OF HOW THE DATA IS USEFUL**
		presented in the training have been learned by the participants. It is more difficult to measure learning than to merely solicit reaction. Measures of learning should be objective, with quantifiable indicators of how new requirements are understood and absorbed. This data is used to confirm that participant learning has occurred as a result of the training initiative. This data also is used to make adjustments in the program content, design, and delivery.
Level 3: Job Application and/or implementation	Focus is on the participant, the work setting, and support mechanisms for applying learning	This evaluation measures behavioral change on the job. It may include specific application of the special knowledge, skills, etc., learned in the training. It is measured after the training has been implemented in the work setting. It may provide data that indicate the frequency and effectiveness of the on-the-job application. It also addresses why the application is or is not working as intended. If it is working, we want to know why, so we can replicate the supporting influences in other situations. If it is not

Table 1.1. Five levels and six types of measures (*continued*).

EVALUATION FRAMEWORK		
LEVEL AND TYPE OF DATA	**FOCUS OF THE DATA**	**SUMMARY OF HOW THE DATA IS USEFUL**
		working, we want to know what prevented it from working so that we can correct the situation in order to facilitate other implementations.
Level 4: Business impact	Focus is on the impact of the training process on specific organizational outcomes	A training program often is initiated because one or more business measures is below expectation or because certain factors threaten an organization's ability to perform and meet goals. This evaluation determines the training's influence or impact in improving organizational performance. It often yields objective data such as costs savings, output increases, time savings, or quality improvements. It also yields subjective data, such as increases in customer satisfaction or employee satisfaction, customer retention, improvements in response time to customers, etc. Generating business impact data includes collecting data before and after the training and linking the outcomes of the training to the appropriate business measures by analyzing the resulting improvements (or lack thereof) in business performance.

Table 1.1. Five levels and six types of measures (*continued*).

EVALUATION FRAMEWORK		
LEVEL AND TYPE OF DATA	**FOCUS OF THE DATA**	**SUMMARY OF HOW THE DATA IS USEFUL**
Level 5: Return on investment (ROI)	Focus is on the monetary benefits as a result of the training	This is an evaluation of the monetary value of the business impact of the training, compared with the costs of the training. The business impact data is converted to a monetary value in order to apply it to the formula to calculate return on investment. This shows the true value of the program in terms of its contribution to the organization's objectives. It is presented as an ROI value or cost-benefit ratio, usually expressed as a percentage. An improvement in a business impact measure as a result of training may not necessarily produce a positive ROI (e.g., if the training was very expensive).
Intangible benefits	Focus is on the added value of the training in non-monetary terms	Intangible data is data that either cannot or should not be converted to monetary values. This definition has nothing to do with the importance of the data; it addresses the lack of objectivity of the data and the inability to convert the data to monetary values. Sometimes it may be too expensive to convert certain data to a monetary value. Other times,

Table 1.1. Five levels and six types of measures (*continued*).

EVALUATION FRAMEWORK		
LEVEL AND TYPE OF DATA	FOCUS OF THE DATA	SUMMARY OF HOW THE DATA IS USEFUL
		management and other stakeholders may be satisfied with intangible data. Subjective data that emerge in evaluation of business impact may fall into this category (e.g., increases in customer satisfaction or employee satisfaction, customer retention, improvements in response time to customers). Other benefits that are potentially intangible are increased organizational commitment, improved teamwork, improved customer service, reduced conflicts, and reduced stress. Often, the data tell us that such things have been influenced in a positive way by the training (so there presumably is a business impact), but the organization has no monetary way to measure the impact. A business impact that cannot be measured in monetary terms cannot be compared with the cost of the training, so no cost-benefit ratio, or ROI, can be determined. This places the data in the intangible category.

The six types of data that are the focal point of this book are all useful in their own ways and for specific purposes. A chapter in this book is dedicated to each of the six types of data. Each chapter presents the merits of the specific type of data and how it is used.

When planning an evaluation strategy for a specific program, an early determination must be made regarding the level of evaluation to be used. This decision is always presented to interested stakeholders for their input and guidance. For example, if you have decided to evaluate a specific program (or a stakeholder has asked for an evaluation), you should first decide the highest level of evaluation that is appropriate. This should guide you as to the purpose of the evaluation study. You should then ascertain what data is acceptable to the various stakeholders and what interests and expectations they have for each of the five levels. After an appropriate discussion about possible intangibles, you should seek their opinions and expectations about the inclusion of intangible data.

The five levels

The levels represent the first five of the six measures (key indicators) discussed in this book. At Level 1, Reaction, Satisfaction, and Planned Actions, participants' reactions to the training are measured, along with their input on a variety of issues related to training design and delivery. Most training programs are evaluated at Level 1, usually by means of generic questionnaires or surveys.

Although this level of evaluation is important as a measure of customer satisfaction, a favorable reaction does not ensure that participants have learned the desired facts, skills, etc., will be able to implement them on the job, and/or will be supported in implementing them on the job. An element that adds value to a Level-1 evaluation is to ask participants how they plan to apply what they have learned.

At Level 2, Learning, measurements focus on what the participants learned during the training. This evaluation is helpful in determining whether participants have absorbed new knowledge and skills and know how to use them as a result of the training. This is a measure of the success of the training program. However, a positive measure at this level is no guarantee that the training will be successfully applied in the work setting.

At Level 3, Application and Implementation, a variety of followup methods are used to determine whether participants actually apply what they have learned from the training to their work settings. The frequency and effectiveness of their use of new skills are important measures at Level 3. Although Level-3 evaluation is important in determining the application of the training, it still does not guarantee that there will be a positive impact on the organization.

At Level 4, Business Impact, measurement focuses on the actual business results achieved as a consequence of applying the knowledge and skills from the training. Typical Level-4 measures are output, quality, cost, time, and customer satisfaction. However, although the training may produce a positive measurable business impact, there is still the question of whether the training may have cost too much, compared to what it achieved.

At Level 5, Return on Investment, the measurement compares the monetary value of the benefits resulting from the training with the actual costs of the training program. Although the ROI can be expressed in several ways, it usually is presented as a percentage or benefit-cost ratio. The evaluation cycle is not complete until the Level-5 evaluation has been conducted.

The "chain of impact" means that participants learn something from the training that they apply on the job (new behavior) that produces an impact on business results (Level 4). Figure 1.1 illustrates the chain of impact between the levels and the value of the information provided, along with frequency and difficulty of assessment. As illustrated on the left side of the figure, for training to produce measurable business results, the Chain of Impact must occur. In evaluating training programs, evidence of results must be collected at each level up to the top one that is included, in order to determine that this link-age exists. For example, if Level 3 will be the highest level evaluated, then data must be collected at Level 3 and Level 2 to show the chain of impact, but it is not necessary to collect Level-4 and -5 data. Although Level-1 data is desirable, it is not always necessary in order to show linkage. However, when possible, L-1 data should be collect-ed as an additional source of information. As shown in Figure 1.1, sim-ple evaluations, such as Level-1 Reactions, are done more frequently than are evaluations at higher levels, which involve more complexity.

Interest in the different levels of data varies, depending on the requirements of the stakeholder. As illustrated, clients (stakeholders who fund training initiatives) are more interested in business impact and ROI data, whereas consumers (participants) are more interested in reaction, learning, and perhaps application. Supervisors and/or team leaders who influence participation in training are often more interested in application of learning in the work setting.

Figure 1.1. Characteristics of evaluation levels.

CHAIN OF IMPACT	VALUE OF INFORMATION	CUSTOMER FOCUS	FREQUENCY OF USE	DIFFICULTY OF ASSESSMENT
1. Reaction	Lowest	Consumer	Frequent	Easy
2. Learning				
3. Application				
4. Impact				
5. ROI	Highest	Client	Infrequent	Difficult

Customers: *Consumers are customers (participants) who are actively involved in the training.*

Clients are customers (stakeholders) who fund, support, and approve the training.

CASE ILLUSTRATION: UTILITY SERVICES COMPANY

A case presentation, "Utility Services Company," helps to illustrate the value of the six different types of data. It presents a training scenario and builds on levels of success that demonstrate increasing importance to the organization (ultimately, return on investment). Following the case illustration, the relative value of each level of data is presented for your review.

Utility services company

PROGRAM SUCCESS IS REPORTED IN A VARIETY OF WAYS. WHICH WOULD YOU PREFER TO RECEIVE?

The program

A team-building program was conducted with 18 team leaders in the operations areas of a water, gas, and electricity services company. For each team, a variety of quality, productivity, and efficiency measures were routinely tracked to reflect team performance. The program was designed to build five essential core skills needed to energize the team to improve team performance. Productivity, quality, and efficiency measures should improve with the application of team-leadership skills. The program consists of three days of classroom learning with some limited follow-up. Experiential exercises were used in most of the team-building processes. The program manager was asked to report on the success of the program. The following options are available:

The results (option A)

1. Program feedback was very positive. Program participants rated the course 4.2 out of 5 in an overall assessment. Participants enjoyed the program and indicated that it was relevant to their jobs. Sixteen participants planned specific activities to focus on team building on the job.

The results (option B)

1. Program feedback was very positive. Program participants rated the course 4.2 out of 5 in an overall assessment. Participants enjoyed the program and indicated that it was relevant to their jobs. Sixteen participants planned specific activities to focus on team building on the job.

2. Participants learned new team-leadership skills. An observation of skill practices verified that the team members acquired adequate skills in the five core team-leadership skills. In a multiple-choice, self-scoring test on team building and team motivation, a 48% improvement was realized when comparing pre- and post scores.

The results (option C)

1. Program feedback was very positive. Program participants rated the course 4.2 out of 5 in an overall assessment. Participants enjoyed the program and indicated that it was relevant to their jobs. Sixteen participants planned specific activities to focus on team building on the job.

2. Participants learned new team-leadership skills. An observation of skill practices verified that the team members acquired adequate skills in the five core team-leadership skills. In a multiple choice, self-scoring test on team building and team motivation, a 48% improvement was realized when comparing pre- and post scores.

3. Participants applied the skills on the job. On a follow-up questionnaire, team leaders reported high levels of use of the five core team-leadership skills learned from the program. In addition, participants identified several barriers to the transfer of skills into actual job performance.

The results (Option d)

1. Program feedback was very positive. Program participants rated the course 4.2 out of 5 in an overall assessment. Participants enjoyed the program and indicated that it was relevant to their jobs. Sixteen participants planned specific activities to focus on team building on the job.

2. Participants learned new team-leadership skills. An observation of skill practices verified that the team members acquired adequate skills in the five core team-leadership skills. In a multiple choice, self-scoring test on team building and team motivation, a 48% improvement was realized when comparing pre- and post scores.

3. Participants applied the skills on the job. On a follow-up questionnaire, team leaders reported high levels of use of the five core team leadership skills learned from the program. In addition, participants identified several barriers to the transfer of skills into actual job performance.

4. Performance records from teams units reflect the following improvements in the sixth month following completion of the program: productivity has improved 23%, combined quality measures have improved 18%, and efficiency has improved 14.5%. While other factors have influenced these measures, the program designers feel that the team-building program had an important impact on these business measures. The specific amount cannot be determined.

The results (option E)

1. Program feedback was very positive. Program participants rated the course 4.2 out of 5 in an overall assessment. Participants enjoyed the program and indicated that it was relevant to their jobs. Sixteen participants planned specific activities to focus on team building on the job.

2. Participants learned new team-leadership skills. An observation of skill practices verified that the team members acquired adequate skills in the five core team-leadership skills. In a multiple choice, self-scoring test on team building and team motivation, a 48% improvement was realized when comparing pre- and post scores.
3. Participants applied the skills on the job. On a follow-up questionnaire, team leaders reported high levels of use of the five core team-leadership skills learned from the program. In addition, participants identified several barriers to the transfer of skills into actual job performance.
4. Performance records from team's units reflect the following improvements in the sixth months following completion of the program: productivity has improved 23%, combined quality measures have improved 18%, and efficiency has improved 14.5%.

Several other factors were identified which influenced the business impact measures. Two other initiatives helped improve quality. Three other factors helped to enhance productivity, and one other factor improved efficiency. Team leaders allocated the percentage of improvement to each of the factors, including the team-building program. To accomplish this, team leaders were asked to consider the connection between the various influences and the resulting performance of their teams and indicate the relative contribution of each of the factors. The values for the contribution of the team-building program are presented below. Because this is an estimate, a confidence value was placed on each factor, with 100% representing certainty and 0 representing no confidence. The confidence percentage is used to adjust the estimate. This approach adjusts for the error of the uncertainty of this estimated value.

The adjustments are shown below:

	MONTHLY IMPROVEMENT IN SIX MONTHS A	PERCENT CONTRIBUTION FROM TEAM BUILDING B	AVERAGE CONFIDENCE ESTIMATE (PERCENT) C	ADJUSTED IMPROVEMENT IN SIX MONTHS A x B x C
Productivity	23%	57%	86%	11.3%
Quality	18%	38%	74%	5%
Efficiencies	14.5%	64%	91%	8.4%

The results (option F)

The data in Option E are developed plus costs and values. Recognizing that the cost of the program might exceed the benefits, the program manager developed the fully loaded cost for the team-building program and compared it directly with the monetary benefits. This required converting the productivity, quality, and efficiency measures to monetary amounts using standard values available in the work units. The benefits are compared directly to the program costs using an ROI formula. Calculations are as follows:

Program Costs for Eighteen Participants = $54,300
Annualized First-Year Benefits

Productivity	197,000
Quality	121,500
Efficiency	90,000
	$408,500 Total Program Benefits

$$\text{ROI} = \frac{408,500 - 54,300}{54,300} \times 100 = 652\%$$

Table 1.2. Relative value of data.

ADDITIONAL EVALUATION DATA PROVIDED	RELATIVE VALUE OF THE DATA
Option A. Program feedback about the participants	We know what the participants think about the program and the job-application possibilities.
Option B. Participants learned leadership skills	We know that the learning new team-occurred.
Option C. Participants applied team-leadership skills on the job.	We know that the skills are being used in the work setting and we know what some of the barriers are that may be preventing optimum utilization and performance.
Option D. Performance records from team's units reflect improvements in the sixth month following completion of the program: productivity has improved 23%, combined quality measures have improved 18%, and efficiency has improved 14.5%.	We know that each of these three measures improved. We are uncertain if the improvements are linked to the training at all. The data shows a definite improvement (23%, 18%, and 14.5%), but since no value has been placed on the measures, it is intangible data.
Option E. Several other factors were identified that influenced the business impact measures. Team leaders allocated the percentage of improvement to each of the factors, including the team-building program.	We have an estimate (adjusted for error) as to why the measures improved and the extent to which the training influenced the improvements. The data shows a definite improvement linked to the training, but since no value has been placed on the measures, it is intangible data.

Table 1.2. Relative value of data (*continued*).

ADDITIONAL EVALUATION DATA PROVIDED	RELATIVE VALUE OF THE DATA
Option F. Fully loaded costs for the team-building program are compared directly with the monetary benefits, using standard values for productivity, quality, and efficiency in the work units	We know that the benefits of the training exceeded the fully loaded cost of the training. We also have an estimate as to how much (652%).

RELATIVE VALUE OF DATA

Each version of the data from the Utility Services Company case has relative value to the organization as the level of information is developed. Table 1.2 illustrates this relative value.

The ROI Process creates a balanced evaluation by collecting, measuring and reporting six types of data:

1. Reaction to/satisfaction with the training and planned actions.
2. Learning.
3. Application/implementation on the job.
4. Business impact.
5. Return on investment (financial impact).
6. Intangible benefits.

This allows for the contribution of the training to be presented in context and in a credible manner. It also accommodates the presentation of the type of data in which each stakeholder has a stated interest.

SETTING EVALUATION TARGETS

Because evaluation processes are constrained by budgets and other resources, it is both useful and prudent to evaluate an organization's training programs by using sampling techniques, with different levels of evaluation being conducted according to predeter-

Figure 1.2. Setting evaluation targets.

EVALUATION LEVEL	TARGET
Level 1, Reaction/Satisfaction Planned Action	100%
Level 2, Learning	50%
Level 3, Job Application	30%
Level 4, Business Impact	20%
Level 5, ROI	10%

mined percentages. An example used by a large telecommunications company, presented in Figure 1.2, illustrates how this works.

As an example, if 100 programs are to be delivered during the year, all of them will be evaluated at Level 1 and 10 of them will be evaluated at Level 5. Since a Level-5 evaluation also includes all other levels, the 10 programs evaluated at Level 5 would be included in the evaluations at each of the other levels. For example, 10 of the 50 programs being evaluated at Level 2 would also be evaluated at Levels 3, 4, and 5 as part of the ROI target.

These targets represent what a large company with a dedicated evaluation group would pursue. This may be too aggressive for some organizations. Targets should be realistically established, given the resources available. Discuss this with your training colleagues and stakeholders and decide how to set similar targets for evaluation in your organization.

Evaluation targets for your organization.

Level 1   _____%

Level 2   _____%

Level 3   _____%

Level 4   _____%

Level 5   _____%

CREATING A RESULTS-BASED TRAINING CULTURE

In order for training to achieve optimum results and sustain credibility in an organization, it is important that a culture of results-based training exist in the organization. Table 1.3 illustrates the fundamentals of this culture.

The ROI Process presented in the succeeding chapters will guide you in planning and implementing evaluation of training and development programs, whether you and your stakeholders decide to evaluate only through, for example, Level 3, or to utilize the entire process and generate all six types of data.

FURTHER READING

[1]Kirkpatrick, Donald L. *Evaluating Training Programs: The Four Levels*, 2nd Edition. San Francisco: Berrett-Koehler Publishers, 1998.

Table 1.3. (Downloadable Form.) A results-based training culture.

ORGANIZATIONAL CHARACTERISTIC	WHAT IT MEANS
The programs are initiated, developed, and delivered with the end result in mind.	*The program objectives are stated not only in terms of learning, but also what the participant is expected to do in the work setting and the impact it should have on business performance, expressed (if possible) in measurable terms.*
A comprehensive measurement and evaluation system is in place for each training program.	*Measurements are defined when training programs are designed or purchased.*
Level 3, 4, and 5 evaluations are regularly developed.	*Throughout the training function, some programs are evaluated for application in the work setting, business impact, and ROI.*
Program participants undestand their responsibility to obtain results as a result of the programs.	*Participants understand what is expected from them as a result of each program, even before they participate. They expect to be held accountable for learning and for applying what they learn.*
Training support groups (management, supervisors, co-workers, etc.) help to achieve results from training.	*All stakeholders, and particularly immediate supervisors/managers and team members, carry out their responsibilities in creating a performance culture that initiates and continues the learning process.*

2

The ROI Model and Process

OVERVIEW OF THE ROI MODEL AND PROCESS

The ROI Process has been used in hundreds of business and government organizations to demonstrate the impact and return on investment of training programs, human resource programs, major change initiatives, and performance-improvement programs. The four features valued most by clients are simplicity, cost-effectiveness, flexibility, and robust feedback useful for informing senior management about performance on the job and impact on business measures. Figure 2.1 illustrates each component of the ROI Process. Each component is given a number to aid in briefly describing it. The description follows Figure 2.1.

Develop Objectives of Training (#1): This initial step develops an understanding of the scope of the program and the business measures that it should influence. If the program is an existing program being evaluated, the objectives and content of the program are reviewed to guide the development of evaluation strategies. If it is a new program, needs assessment data are used to develop objectives at levels 1 through 4. The purpose of the evaluation study is then determined.

Develop Evaluation Plans and Baseline Data (#2): The Data Collection Plan is developed and measurements (4 levels), methods

Figure 2.1. The ROI model and process.

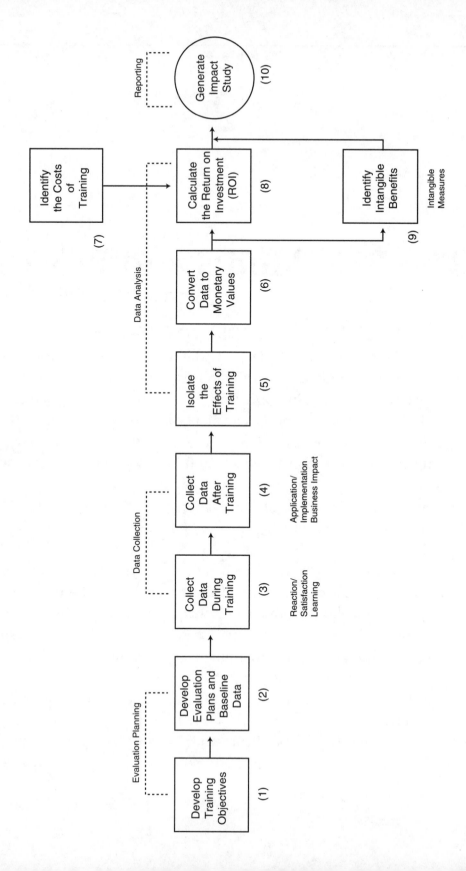

of data collection, sources of data, and timing of collection are identified to collect baseline and follow-up data. The ROI Analysis Plan is developed and the methods of isolation, conversion of data to monetary values, cost categories, communication targets, and other steps are determined. The nature of the training intervention and the roll-out schedule will dictate the timing of the data gathering. The purpose of the study and appropriate evaluation strategies are verified before beginning the process. The completion of these plans completes the planning process for the remaining steps (3 through 10).

Collect Data During Training (#3): The training is implemented, and data is collected at level 1 and level 2. The evaluator may not always be involved in collecting data at these two levels, but should require evidence from others (especially at level 2) that provides sufficient data to satisfy the needs of the study at the level in question.

Collect Follow-up Data After Training (#4): Applying the methods and timing from the Data Collection Plan described earlier, follow-up data is collected. Depending on the program selected for evaluation as described in the Data Collection Plan, data collection may utilize questionnaires, interviews, data from company records, or other methods as appropriate. The cost of the training (#7) is tabulated per the guidelines on the ROI Analysis Plan and will be used later in the ROI calculation.

Isolate the Effects of the Training (#5): As indicated on the ROI Analysis Plan, one or more strategies are used to isolate the effects of the training. Examples are use of a control group arrangement, trend line analysis, estimates by participants, estimates by managers, and estimates by in-house experts. If a control group arrangement is feasible, performance data will be collected on the trained group and on another group with similar characteristics that does not receive the training. The pre- and post-training performance of the two groups will be compared to determine the extent of improvement influenced by the training. At least one isolation strategy will be used to determine the extent of influence the training intervention has on key business measures.

Convert Data to Monetary Values (#6): Certain business impact data influenced by the training will be converted to monetary values. This is necessary in order to compare training benefits to training costs to determine the return on investment (calculate the ROI, #8). Fully loaded costs (#7) must be captured in order to complete the calculation. If some data cannot be converted to a monetary value, that data will be reported either as business impact results (e.g., improvements in customer or employee satisfaction) or as intangible benefits when the business impact cannot be expressed as a hard value (# 9).

Generate an Impact Study (#10): At the conclusion of the study, two reports are usually developed for presentation. One report is brief and intended for presentation to executive management. The other report is more detailed and is suitable for other stakeholders.

DEFINING THE RETURN ON INVESTMENT AND BENEFIT-COST RATIO
Basic Concept of ROI

The term return on investment is often misused in the training and performance improvement field, sometimes intentionally. In some situations, a very broad definition for ROI is used to include any benefit from a training intervention. In these situations, ROI is a vague concept where even subjective data linked to an intervention are included in the concept of the return. In the ROI Process presented in this book, the return on investment is more precise and is meant to represent an actual value developed by comparing training-intervention costs to outcome benefits. The two most common measures are the benefit-cost ratio and the ROI formula. Both are presented in this chapter.

Annualized Values

All the data presented in this book use annualized values so that the first-year impact of the investment in a training program is developed. Using annual values is becoming a generally accepted practice for developing the ROI in many organizations. This approach is a conservative way to develop the ROI, since many short-term training and performance improvement initiatives have added value in the second or third year. For long-term interventions, annualized values are inappropriate and longer time frames need to be used. For most training interventions of one-day to one-month duration, first-year values are appropriate.

Benefit-Cost Ratio

One method for evaluating training and performance improvement investments compares the annual economic benefits of a training intervention to its costs, using a ratio. In formula form, the ratio is:

$$BCR = \frac{\text{Training Benefits}}{\text{Training Costs}}$$

A benefit-cost ratio of 1 means that the benefits equal the costs. A benefit-cost ratio of 2, usually written as 2:1, indicates that for each dollar spent on the training, two dollars were returned as benefits.

The following example illustrates the use of the benefit-cost ratio. An applied leadership-training program, designed for managers and supervisors, was implemented at an electric and gas utility. In a follow-up evaluation, action planning and business performance monitoring were used to determine benefits. The first-year payoff for the intervention was $1,077,750. The total, fully loaded implementation costs were $215,500. Thus, the benefit-cost ratio was:

$$BCR = \frac{\$1,077,750}{\$215,500} = 5$$

This is expressed as 5:1, meaning that for every one dollar invested in the leadership program, five dollars in benefits is returned.

The ROI formula

Perhaps the most appropriate formula to evaluate training and performance improvement investments is to use net benefits divided by cost. The ratio is usually expressed as a percentage when the fractional values are multiplied by 100. In formula form, the ROI becomes:

$$\text{ROI} = \frac{\text{Net Training Benefits}}{\text{Training Costs}} \times 100 = \underline{\quad}\%$$

Net benefits are training benefits minus training costs. The ROI value is related to the BCR by a factor of one. For example, a BCR of 2.45 is the same as an ROI value of 145%. This formula is essentially the same as ROI in other types of investments. For example, when a firm builds a new plant, the ROI is annual earnings divided by investment. The annual earnings are comparable to net benefits (annual benefits minus the cost). The investment is comparable to the training costs, which represent the investment in the training program.

An ROI on a training investment of 50% means that the costs are recovered and an additional 50% of the costs is reported as "earnings." A training investment of 150% indicates that the costs have been recovered and an additional 1.5 times the costs is captured as "earnings." An example is provided below using the same leadership program and results illustrated for the BCR above.

$$\text{ROI} (\%) = \frac{\$1,077,750 - \$215,500}{\$215,500} \times 100 = 400\%$$

For each dollar invested, four dollars were received in return, after the cost of the program had been recovered. Using the ROI formula essentially places training investments on a level playing field with other investments using the same formula and similar concepts. The ROI calculation is easily understood by key management and financial executives who regularly use ROI with other investments.

Although there are no generally accepted standards, some organizations establish a minimum requirement or hurdle rate for an ROI in a training or performance improvement initiative. An ROI minimum of 25% is set by some organizations. This target value is usually above the percentage required for other types of investments. The rationale is that the ROI process for training is still a relatively new concept and often involves subjective input, including estimations. Because of that, a higher standard is required or suggested. Target options are listed below.

- Set the value as with other investments, e.g. 15%
- Set slightly above other investments, e.g. 25%
- Set at break even (0%)
- Set at client expectations

DECIDING WHICH OF THE FIVE LEVELS IS RIGHT FOR YOUR TRAINING EVALUATION

Evaluation dollars must be spent wisely. As mentioned previously, sampling can be used extensively to gather data and get a good picture of how training is making a contribution to the organization. Since level-4 and level-5 evaluations consume the most resources, it is suggested that evaluations at this level be reserved for programs that meet one or more of the following criteria:

- The life cycle of the program is such that it is expected to be effective for at least 12 to 18 months.
- The program is important in implementing the organization's strategies or meeting the organization's goals.
- The cost of the program is in the upper 20 percent of the training budget.

- The program has a large target audience.
- The program is highly visible.
- Management has expressed an interest in the program.

Level-3 evaluations are often prescribed for programs that address the needs of those who must work directly with customers, such as sales representatives, customer-service representatives, and those in call centers who must engage in customer transactions immediately after the training program. Compliance programs are also good candidates for Level-3 evaluation. (See Figure 2.2.)

The worksheet in Figure 2.2 should be used as a guide as you review your curriculum and decide which programs to evaluate at each level. Since your evaluation resources are scarce, this will be useful to help narrow your choices.

The upper portion of the worksheet is used to narrow your choices in determining which programs are the best candidates for Level-4 and -5 evaluation. Once you have ranked the possibilities, you will still need to make the final decision based on budget and other resources. It is best that a team of people utilize the worksheet to process these decisions. Designers, instructors, and managers familiar with the programs can serve on the team.

The lower portion of the worksheet is used to guide your decisions on Level 3 candidates. The criteria for Level 3 is not as strict as the Level-4 and -5 criteria. Your decisions for Level 3 should be based more on visible impact on customers, revenue, and the implications of proper employee behavior, as expected to be influenced by a particular program.

Programs that do not meet the criteria for Level 3, 4, or 5 evaluation, should be considered for Level-1 and/or Level-2 evaluation based on the objectives of the program, the expected cost of the evaluation, the ease of evaluation, and the value of the data derived from the evaluation. For example, Level-1 evaluation is inexpensive, easy to do, and yields useful data that can improve a program. Level-2

Figure 2.2. Downloadable worksheet: Selecting programs for evaluation at each of the 5 levels.

Worksheet—Program Selection Criteria: Selecting Programs/Interventions for Level-4 and -5 Evaluation

Step One: List ten of your programs below above columns 1 through 10.

Step two: Use the 1-through-5 rating scale to rate each program on each Level-4 and Level-5 evaluation criteria A through F.

Step three: Use the total scores (A through F) to determine which programs to evaluate at Level-4 and 5 this budget year.

Program/Intervention →

L-4 and 5 Evaluation Criteria	1	2	3	4	5	6	7	8	9	10
A. Program Life Cycle										
B. Strategic Objectives/Goals										
C. Cost of Program										
D. Audience Size Over Life of Program										
E. Visibility of Program										
F. Management Interest										
Total Score A through F										

Level- and -5 Criteria   *Rating Scale for Level-4 and -5 Decisions*

A. Program Life Cycle   5 = Life cycle viable 12 to 18 months or longer (permanent program); 1 = Very short life cycle (one shot program)

B. Organizations Strategic Objectives/Goals   5 = Closely related to implementing organization's strategic objectives or goals; 1 = Not related to strategic objectives/goals

C. Cost of Program, Including Participants Sal/Benefits   5 = Very expensive; cost of the program is in the upper 20 percent of the training budget; 1 = Very inexpensive

D. Audience Size Over Life of Program   5 = Very large target audience, 1 = Very small target audience

E. Visibility of Program   5 = High visibility for program with significant stakeholder(s); 1 = Low visibility for program with significant stake holder(s)

F. Management Interest   5 = High level of management interest in evaluating this program; 1 = Low level of management interest in evaluating program.

Figure 2.2. Downloadable worksheet: Selecting programs for evaluation at each of the 5 levels (*continued*)

Worksheet-Program Selection Criteria: Selecting Programs/Interventions for Level-3 Evaluation

List your programs that meet the criteria below. Determine which ones merit being evaluated at Level 3 during this budget year by circling Yes or No.

List names of your programs ⟶

Level-3 Criteria	Program 1		Program 2		Program 3		Program 4		Program 5		Program 6	
	Level-3 Rating Circle Yes or No ⟶											
Compliance Program	Yes	No	Yes	No	Yes	No	Yes	No	Yes	No	Yes	No
Customer Service Program	Yes	No	Yes	No	Yes	No	Yes	No	Yes	No	Yes	No
Sales Program	Yes	No	Yes	No	Yes	No	Yes	No	Yes	No	Yes	No
Call Center or Other Customer Transaction Program	Yes	No	Yes	No	Yes	No	Yes	No	Yes	No	Yes	No
Organization Sponsored Certification Program	Yes	No	Yes	No	Yes	No	Yes	No	Yes	No	Yes	No

Note: Any program may be reviewed for Level-4 and-5 evaluation. Additionally, a program qualifying for L-3 evaluation (receives a "yes" response) should also be reviewed for possible Level-4 and-5 evalution. An exception is "Compliance Programs," which except for rare occasions/circumstances, are not candidates for evaluation above Level 3.

evaluation is likely a requirement for certification programs or programs that address safety issues or customer service issues.

Evaluation decisions must also be made based on capabilities and resource availability. In any event, you can be sure that others are evaluating your programs, whether by word of mouth or by personal experience. Unless training practitioners want training programs and the training function to be judged on subjective approaches and hearsay, they must prepare themselves to do the job and they must allocate the resources to do a thorough job. Perhaps we should ask: is it worth 5% of the training budget to determine if the other 95% is expended on programs that are making the proper contribution?

FURTHER READING

Kaufman, Roger, Sivasailam Thiagarajan, and Paula MacGillis, editors. *The Guidebook for Performance Improvement: Working with Individuals and Organizations.* San Francisco: Jossey-Bass/Pfeiffer, 1997.

Kirkpatrick, Donald L. *Evaluating Training Programs: The Four Levels,* 2nd Edition. San Francisco: Berrett-Koehler Publishers, 1998.

Phillips, Jack J. *Handbook of Training Evaluation and Measurement Methods,* 3rd Edition. Houston: Gulf Publishing, 1997.

Swanson, Richard A., and Elwood F. Holton III. *Results: How to Assess Performance, Learning, and Perceptions in Organizations.* San Francisco: Berrett-Koehler Publishers, 1999.

Phillips, Jack J. "Was It The Training?" *Training & Development,* Vol. 50, No. 3, March 1996, pp. 28-32.

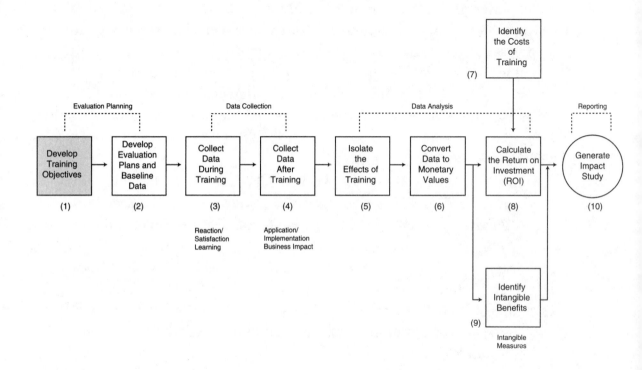

Evaluation Planning

| Develop Training Objectives | Develop Evaluation Plans and Baseline Data | Collect Data During Training | Collect Data After Training | Isolate the Effects of Training | Convert Data to Monetary Values | Calculate the Return on Investment (ROI) | Generate Impact Study |
| (1) | (2) | (3) | (4) | (5) | (6) | (8) | (10) |

Identify the Costs of Training
(7)

Data Collection

Data Analysis

Reporting

Reaction/
Satisfaction
Learning

Application/
Implementation
Business Impact

Identify
Intangible
Benefits
(9)

Intangible
Measures

3

Step 1. Develop Training Objectives: The Basis for Measurement

HOW SPECIFIC OBJECTIVES AT EACH LEVEL CONTRIBUTE TO RETURN ON INVESTMENT

Developing specific objectives for training programs at each of the five levels provides important benefits. First, it provides direction to the program designers, analysts, and trainers directly involved in the training process to help keep them on track. Objectives define exactly what is expected at different time frames from different individuals and involving different types of data. They also provide guidance as to the outcome expected from the training and can serve as important communication tools to the support staff, clients, and other stakeholders so that they fully understand the ultimate goals and impact of the training. For example, Level-3-and-4 objectives can be used by supervisors of eligible employees and by prospective participants to aid in the program selection process. Participants can translate learning into action when they know the linkage between L-2, L-3, and L-4. Finally, from an evaluation perspective, the objectives provide a basis for measuring success at each of the levels.

LINKING TRAINING OBJECTIVES TO ORGANIZATIONAL OBJECTIVES

The results-based approach in Figure 3.1 illustrates how evaluation spans the complete training process. Phase A, the top row, depicts the process of identifying the needs at each level of the framework (Levels 1 through 5). The result of this phase of the training process is the identification of the specific problem/opportunity, the identification of the performance gap and why it exits, and identification of the appropriate solution, including training and any other solution(s) to close the gap. The last step of this component, when training is a solution, is to develop the objectives of the training, design the strategy and measures to evaluate the training at each of the five levels, and complete the additional planning that precedes the implementation of the evaluation process.

Phase B, the middle row, begins with the design or purchase of the solution and ends with implementation of the solution. It is at this point that Level-1 and Level-2 evaluation is implemented and data are collected per the evaluation strategy and plans that were previously developed.

Phase C, the bottom row, is the follow-up evaluation component. The training has been delivered and sufficient time has passed to allow us to determine if the training is working and the extent to which organizational impact has occurred. When the decision is made to conduct a follow-up evaluation of a training program, the planning is already in place. The plan can be implemented and the data collected during the time frame identified by the plan. Then the data are analyzed, the training is isolated, data are converted to monetary values, and the ROI is calculated. The intangible benefits are identified and presented in a report along with the ROI calculation and supporting data and recommendations.

Achieving the best results requires that training needs be properly identified and that relevant objectives be developed. Developing evaluation plans at the time of needs assessment can greatly influence the design, delivery, and outcome of the training.

Figure 3.1. Results-based approach.

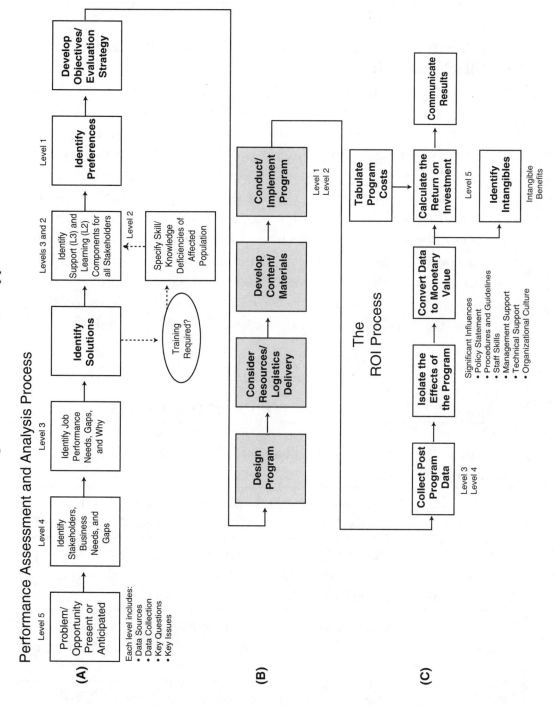

Performance Assessment and Analysis Process

Level 5
Problem/ Opportunity Present or Anticipated

Level 4
Identify Stakeholders, Business Needs, and Gaps

Level 3
Identify Job Performance Needs, Gaps, and Why

Identify Solutions

Levels 3 and 2
Identify Support (L3) and Learning (L2) Components for all Stakeholders

Level 1
Identify Preferences

Develop Objectives/ Evaluation Strategy

Training Required?

Level 2
Specify Skill/ Knowledge Deficiencies of Affected Population

Each level includes:
• Data Sources
• Data Collection
• Key Questions
• Key Issues

(A)

Design Program

Consider Resources/ Logistics Delivery

Develop Content/ Materials

Conduct/ Implement Program

Level 1
Level 2

(B)

The ROI Process

Collect Post Program Data

Isolate the Effects of the Program

Convert Data to Monetary Value

Tabulate Program Costs

Calculate the Return on Investment

Communicate Results

Identify Intangibles

Level 3
Level 4

Level 5

Intangible Benefits

Significant Influences
• Policy Statement
• Procedures and Guidelines
• Staff Skills
• Management Support
• Technical Support
• Organizational Culture

(C)

DEVELOPING OBJECTIVES AT EACH LEVEL FOR TRAINING SOLUTIONS

Training solutions are initiated to provide knowledge and skills that influence changes in behavior and should ultimately result in improving organizational outcomes. The need for training stems from an impending opportunity, problem, or need in the organization. The need can be strategic, operational, or both. Whatever the case, the training solutions that are deemed appropriate for the situation should have multiple levels of objectives. These levels of objectives, ranging from qualitative to quantitative, define precisely what will occur as the training is implemented in the organization. Table 3.1 shows the different levels of objectives. These objectives parallel the levels of evaluation in Table 1.1 in Chapter 1 and are so critical that they need special attention in their development and use.

Table 3.1. Levels of objectives.

LEVEL OF OBJECTIVES	FOCUS OF OBJECTIVES
Level 1 Reaction/ Satisfaction	Defines a specific level of satisfaction and reaction to the training as it is delivered to participants.
Level 2 Learning	Defines specific knowledge and skill(s) to be developed/acquired by training participants.
Level 3 Application/ Implementation	Defines behavior that must change as the knowledge and skills are applied in the work setting following the delivery of the training.
Level 4 Business Impact	Defines the specific business measures that will change or improve as a result of the application of the training.
Level 5 ROI	Defines the specific return on investment from the implementation of the training, comparing costs with benefits.

Reaction/satisfaction objectives

The objective of training at level 1 is for participants to react favorably, or at least not negatively. Ideally, the participants should be satisfied with the training since this offers win-win relationships. It is important to obtain feedback on the relevance, timeliness, thoroughness, and delivery aspects of the training. This type of information should be collected routinely so that feedback can be used to make adjustments or even redesign the training.

Learning objectives

All training programs should have learning objectives. Learning objectives are critical to measuring learning because they communicate expected outcomes from the training and define the desired competence or performance necessary to make the training successful. Learning objectives provide a focus for participants to clearly indicate what they must learn, and they provide a basis for evaluating learning. Table 3.2 serves as an aid to assist in the development of learning objectives.

Application/implementation objectives

Application/implementation objectives define what is expected and—often—to what level of performance, when knowledge and skills learned in the training are actually applied in the work setting. Application/implementation objectives are very similar to learning-level objectives but reflect actual use on the job. Application/implementation objectives are crucial because they describe the expected outcomes in the intermediate area, that is, between the learning of new knowledge, skills, tasks, or procedures and the performance that will be improved (the organizational impact). They provide a basis for the evaluation of on-the-job changes and performance. Table 3.3 serves as a job aid to assist in the development of application/implementation objectives.

Table 3.2. (Downloadable form.)

Job Aid: Developing Learning Objectives
Measuring Knowledge and Skill Enhancement

The best learning objectives:

- Describe behaviors that are observable and measurable
- Are outcome-based, clearly worded, and specific
- Specify what the learner must do (not know or understand) as a result of the training
- Have three components:

1. Performance—what the learner will be able to do at the end of the training
2. Condition—circumstances under which the learner will perform the task
3. Criteria—degree or level of proficiency that is necessary to perform the job

Three types of learning objectives are:

- Awareness—Familiarity with terms, concepts, processes
- Knowledge—General understanding of concepts, processes, etc.
- Performance—Ability to demonstrate the skill (at least at a basic level)

Two examples of Level-2 objectives are provided below.

1. Be able to identify and discuss the six leadership models and theories.
2. Given ten customer contact scenarios, with 100 percent accuracy, and be able to identify which steps of the customer interaction process should be applied.

Table 3.3. (Downloadable form.)

Job Aid: Developing Application/Implementation Objectives Measuring On-the-Job Application of Knowledge and Skills

The best application objectives:

- Identify behaviors that are observable and measurable
- Are outcome-based, clearly worded, and specific
- Specify what the participant will change as a result of the training
- May have three components:

1. Performance—what the participant will have changed/accomplished at a specified follow-up time after training
2. Condition—circumstances under which the participant will perform the task
3. Criteria—degree or level of proficiency with which the task will be performed

Two types of application/implementation objectives are:

- Knowledge based-general use of concepts, processes, etc.
- Behavior based-ability to demonstrate use of the skill (at least at a basic level)

Key questions are:

- What new or improved knowledge will be applied on the job?
- What is the *frequency of skill* application?
- What new *tasks* will be performed?
- What new *steps* will be implemented?
- What new *action items* will be implemented?

Table 3.3. (*continued*)

- What new *procedures* will be implemented?
- What new *guidelines* will be implemented?
- What new *processes* will be implemented?

Two examples of Level-3 objectives are provided below.

1. Apply the appropriate steps of the customer interaction process in every customer contact situation.

2. Identify team members who lack confidence in the customer contact process and coach them in the application of the process.

Impact objectives

The objective of every training program should be improvement in organizational performance. Organizational impact objectives are the key business measures that should be improved when the training is applied in the work setting. Impact objectives are crucial to measuring business performance because they define the ultimate expected outcome from the training. They describe business-unit performance that should be connected to the training initiative. Above all, they place emphasis on achieving bottom-line results that key stakeholder groups expect and demand. They provide a basis for measuring the consequences of application (L-3) of skills and knowledge learned (L-2). Table 3.4 serves as a job aid to assist in the development of impact objectives.

Table 3.4. (Downloadable form.)

Job Aid: Developing Impact Objectives
Measuring Business Impact from Application of Knowledge and Skills

The best impact objectives:

- Contain measures that are linked to the knowledge and skills taught in the training program

- Describe measures that are easily collected

- Are results-based, clearly worded, and specific

- Specify what the participant will accomplish in the business unit as a result of the training

Four categories of impact objectives for hard data are:

- Output
- Quality
- Costs
- Time

Three common categories of impact objectives for soft data are:

- Customer service (responsiveness, on-time delivery, thoroughness, etc.)

- Work climate (employee retention, employee complaints, grievances, etc.)

- Work habits (tardiness, absenteeism, safety violations, etc.)

Two examples of Level-4 impact objectives are provided below.

1. Reduce employee turnover from an average annual rate of 25% to an industry average of 18% in one year.

2. Reduce absenteeism from a weekly average of 5% to 3% in six months.

Return-on-investment (ROI) objectives

A fifth level of objectives is the expected return on investment. ROI objectives define the expected payoff from the training and compare the input resources, the cost of the training, with the value of the ultimate outcome—the monetary benefits. An ROI of 0% indicates a break-even training solution. A 50% ROI indicates that the cost of the training is recaptured and an additional 50% in "earnings" is achieved.

For many training programs, the ROI objective (or hurdle rate) is larger than what might be expected from the ROI of other expenditures, such as the purchase of a new company, a new building, or major equipment; but the two are related. In many organizations the ROI objective for training is set slightly higher than the ROI expected from other interventions because of the relative newness of applying the ROI concept to training initiatives. For example, if the expected ROI from the purchase of a new company is 20%, the ROI from a training initiative might be in the 25% range. The important point is that the ROI objective should be established up front and in discussions with the client. The worksheet at the end of this chapter provides more information on this issue.

When the objectives are defined up front, as the needs are identified and the program is designed, it becomes much easier to target results and to develop evaluation strategies for the training. The case illustration, "Reliance Insurance Company," shows how ignoring the identification of needs and objectives results in a lack of benefits and wasted resources.

CASE ILLUSTRATION: RELIANCE INSURANCE COMPANY

At the end of a monthly staff meeting, Frank Thomas, CEO of Reliance Insurance Company, asked Marge Thompson, Manager of Training and Development, about the Communications Workshops that had been conducted with all supervisors and managers throughout the company. The workshop featured the Myers-Briggs Type Indicator® (MBTI) and showed participants how they interact with, and can better understand, one another in their routine activities. The MBTI® instrument classifies people from a range of 16 personality types.

Frank continued, "I found the workshop very interesting and intriguing. I can certainly identify with my particular personality type, but I am curious what specific value these workshops have brought to the company. Do you have any way of showing the results of all 25 workshops?" Marge quickly replied, "We certainly have improved teamwork and communications throughout the company. I hear people make comments about how useful the process has been to them personally." Frank added, "Do we have anything more precise? Also, do you know how much we spent on these workshops?" Marge quickly responded by saying, "I am not sure that we have any precise data and I am not sure exactly how much money we spent, but I can certainly find out." Frank concluded with some encouragement, "Any specifics would be helpful. Please understand that I am not opposing this training effort. However, when we initiate these types of programs, we need to make sure that they are adding value to the company's bottom line. Let me know your thoughts on this issue in about two weeks."

Marge was a little concerned about the CEO's comments, particularly since he enjoyed the workshop and had made several positive comments about it. Why was he questioning the effectiveness of it? Why was he concerned about the costs?

These questions began to frustrate Marge as she reflected over the year-and-a-half period in which every manager and supervisor had attended the workshop. She recalled how she was first introduced to the MBTI®. She attended a workshop conducted by a friend, was impressed with the instrument, and found it to be helpful as she learned more about her own personality type.

Marge thought the process would be useful to Reliance managers and asked the consultant to conduct a session internally with a group of middle-level managers. With a favorable reaction, she decided to conduct a second session with the top executives, including Frank Thomas. Their reaction was favorable. Then she launched it with the entire staff, using positive comments from the CEO, and the feedback had been excellent.

She realized that the workshops had been expensive because over 600 managers had attended. However, she thought that teamwork had improved, although there was no way of knowing for sure. With some types of training you never know if it works, she thought. Still, Marge was facing a dilemma. Should she respond to the CEO or just ignore the issue?

This situation at Reliance Insurance is far too typical. The training needs and objectives were not properly identified, and resources were committed to a project where the benefit to the organization was questionable. This situation can be avoided by properly identifying needs and linking them to appropriate knowledge and skills that will influence the desired outcome.

For example, an assessment of the strength of the training and other factors that may cause a reduction in customer complaints can aid in forecasting the ROI at L-5. *As an example, 200 complaints per month at $300 per complaint = $60,000 monthly cost or $720,000 annually. If the stakeholders believe that the training can influence a reduction of 100 complaints monthly (half), this would save $30,000 or $360,000 annually ($30,000 x 12). By estimating how much we will spend on the training we can determine an expected ROI. If we spend $100,000, our ROI would be 260%. ($360,000 - $100,000 = $260,000/$100,000). Looking at this another way, if our stakeholders will accept an ROI of say, 30%, and we believe our training solution will in fact influence a reduction of complaints by 100, then we can afford to spend up to $277,000 on our training solution ($360,000 - $277,000 = $83,000/$277,000 = 30%).*

Having Level-5 information up front affords many options to stakeholders. When the cost of a problem can be identified in monetary terms, and training has been identified as an appropriate solution, we can get serious about the investment of training resources needed.

Downloadable worksheet—Developing meaningful objectives.

STEPS IN DEVELOPING MEANINGFUL OBJECTIVES WHEN TRAINING IS AN APPROPRIATE SOLUTION	LEVELS 5 AND 4	LEVEL 3	LEVEL 2	LEVEL 1
Step one. Identify stakeholders that have an interest in organizational improvement measures such as improving output, improving quality, time savings, decreasing costs, improving customer satisfaction, improving employee satisfaction, Decreasing complaints, etc.	X			
Step two. Consult with appropriate stakeholders to identify the specific organizational measure(s) that the training is supposed to influence. If possible, identify what the problem is costing the organization. Follow the job aid in Table 3.4 to develop the appropriate Level-4 objectives. *Example: Reduce customer complaints in the customer service call center from 200 per month to less than 50 per month. Excess complaints are costing $300 per complaint on average.*	X			
Step three. Consult with stakeholders to identify the participant behavior that will influence improvement in the specific organizational measures.		X		

Downloadable worksheet—Developing meaningful objectives (*continued*).

STEPS IN DEVELOPING MEANINGFUL OBJECTIVES WHEN TRAINING IS AN APPROPRIATE SOLUTION	LEVELS 5 AND 4	LEVEL 3	LEVEL 2	LEVEL 1
Step four. Follow the job aid in Table 3.3 to develop the appropriate Level-3 objectives. *Example: Apply the appropriate steps of the customer interaction process in every customer contact situation.*		X		
Step five. After developing the appropriate L-3 objectives, identify the knowledge and skill deficiencies that must be addressed by the training to influence the job behaviors. Follow the job aid in Table 3.2 to develop the appropriate Level-2 objectives. *Example: Given 10 customer contact scenarios, with 100 percent accuracy identify which steps of the customer interaction process should be applied.*			X	
Step six. Identify the reaction (L-1) desired from participants when they participate in the training solution. *Example: Overall satisfaction of participants on a 1 to 5 scale, based on course relevance, skill application opportunities, and coaching from facilitator should be at least 4.5*				X

FURTHER READING

Broad, M.L. and Newstrom, J.W., *Transfer of Training*. Reading, MA: Addison-Wesley, 1992.

Kaufman, Roger, Sivasailam Thiagarajan, and Paula MacGillis, editors. *The Guidebook for Performance Improvement: Working with Individuals and Organizations*. San Francisco: Jossey-Bass/Pfeiffer, 1997.

Kirkpatrick, Donald L. *Evaluating Training Programs: The Four Levels*, 2nd Edition. San Francisco: Berrett-Koehler Publishers, 1998.

Phillips, Jack J. *Handbook of Training Evaluation and Measurement Methods*, 3rd Edition. Houston: Gulf Publishing, 1997.

Phillips, Jack J. 1994, 1997. *Measuring the Return on Investment*. Vol 1 and Vol 2. Alexandria, VA: American Society for Training and Development.

Phillips, Jack J. *Return On Investment in Training and Performance Improvement Programs*. Houston, TX: 1997, Gulf Publishing.

Swanson, Richard A., and Elwood F. Holton III. *Results: How to Assess Performance, Learning, and Perceptions in Organizations*. San Francisco: Berrett-Koehler Publishers, Inc. 1999.

Phillips, Jack J. "Was It The Training?" *Training & Development*, Vol. 50, No. 3, March 1996, pp. 28-32.

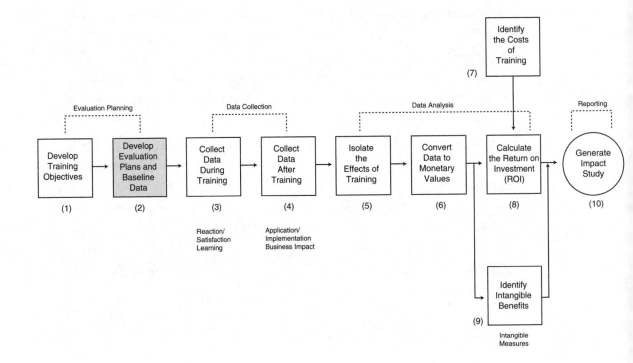

Identify
the Costs
of
Training

(7)

Evaluation Planning   Data Collection   Data Analysis   Reporting

Develop
Training
Objectives

(1)

Develop
Evaluation
Plans and
Baseline
Data

(2)

Collect
Data
During
Training

(3)

Reaction/
Satisfaction
Learning

Collect
Data
After
Training

(4)

Application/
Implementation
Business Impact

Isolate
the
Effects of
Training

(5)

Convert
Data to
Monetary
Values

(6)

Calculate
the Return on
Investment
(ROI)

(8)

Generate
Impact
Study

(10)

Identify
Intangible
Benefits

(9)

Intangible
Measures

4

Step 2. Develop Evaluation Plans and Baseline Data

OVERVIEW OF DEVELOPING EVALUATION PLANS

Evaluation planning is necessary in order to specify, in detail, what the evaluation project will entail. It also specifies and documents how success will be measured. The best time to develop evaluation plans is during the needs-assessment and program-design phases. This is true in part because the need for data and the means to collect and analyze it can be built into the program design. This not only accommodates data collection, but also helps in influencing program designers to focus on measures of success and outcomes. This in turn will influence the facilitators/instructors and the participants as well. Without this focus, training programs often go astray and are implemented with the ultimate expectation being only learning. Developing evaluation plans early in the process also results in a less-expensive evaluation project than what would result from a crisis mode if it is planned shortly before or when the training is implemented. It also allows more time to get the input and concurrence of key stakeholders in the important areas of data collection and analysis. Before discussing the development of evaluation plans further, it is helpful to address the types of measures that are influenced by training.

TYPES OF MEASURES

A fundamental purpose of evaluation is to collect data directly related to the objectives of the training program. As covered in the previous chapter, the needs must be properly identified before the solution can be determined and the objectives developed. As these objectives are developed, they serve as the link between the training solution and the ultimate business outcome. As this link is established, the measures to be influenced by the training solution are identified. Once these measures are identified, it is necessary to determine the current status of the measures so that a baseline can be established. This baseline represents the beginning point when comparing improvements in the measures on a pre- and post-training basis.

In parallel with the development of training objectives, the measures for each objective are identified. If the needs assessment failed to identify the measures properly, then this is another opportunity to ensure alignment between the objectives and the business measures to be influenced. It is helpful at this point to review the types of measures that may be influenced by training. Sometimes the evaluative potential of data is not fully recognized. The confusion sometimes stems from the different types of outcomes planned for training initiatives. Often, training initiatives have skill and behavioral outcomes that promise changes in job-related behavior. The outcome of some training initiatives, such as sales training, are fairly easy to observe and evaluate. However, behavioral outcomes associated with effective management are not nearly so obvious or measurable. Demonstrating that a sales manager is an effective team leader is much more difficult than demonstrating that a sales representative is producing additional sales.

Because of this, a distinction is made between two general categories of data: hard data and soft data. Hard data are measures of improvement, presented as rational, undisputed facts that are easily accumulated. They are the most desired type of data to collect. The ultimate criteria for measuring the effectiveness of training initiatives will rest on hard- and soft-data items. Hard-data items—such as productivity, profitability, cost savings, and quality improvements—are more objective and easier to assign a monetary value. Soft data are more difficult to collect and analyze but are used when hard data are not available.

Hard-data categories

Hard data can usually be grouped in four categories: output, quality, time, and cost. The distinction between these four groups of hard data is sometimes unclear, since there are some overlapping factors. Typical data in these four categories that might be available are shown in Table 4.1.

Soft-data categories

There are situations when hard data are not available or when soft data may be more meaningful to use in evaluating training programs. Soft data can be categorized into six areas: work habits, work climate, attitudes, new skills, development/advancement, and initiative. Table 4.2 includes some typical soft-data items.

There is a place for both hard and soft data in program evaluation. A comprehensive training program will influence both types of data. Hard data are often preferred to demonstrate results because of their distinct advantages and level of credibility. However, soft data can be of equal or greater value to an organization, even though it is more subjective in nature.

CLARIFYING THE PURPOSE OF YOUR EVALUATION INITIATIVE

Evaluation is a systematic process to determine the worth, value, or meaning of an activity or process. In a broad sense, evaluation is undertaken to improve training programs or to decide the future of a training program. The first step in developing an evaluation plan is to define the purpose of the evaluation. This decision drives all other

Table 4.1. Examples of hard data.

OUTPUT	COSTS	TIME	QUALITY
Units produced	Budget variances	Equipment downtime	Scrap
Tons manufactured	Unit costs	Overtime	Waste
Items assembled	Cost by account	On-time shipments	Rejects
Money collected	Variable costs	Time to project completion	Error rates
Items sold	Fixed costs	Processing time	Rework
Forms processed	Overhead cost	Supervisory time	Shortages
Loans approved	Operating costs	Break-in time for new employees	Product defects
Inventory turnover	Number of cost reductions	Training time	Deviation from standard
Patients visited	Project cost savings	Meeting schedules	Product failures
Applications processed	Accident costs	Repair time	Inventory adjustments
Students graduated	Program costs	Efficiency	Time-card corrections
Tasks completed	Sales expense	Work stoppages	Percent of tasks completed properly
Output per hour		Order response	Number of accidents
Productivity		Late reporting	
Work backlog		Lost-time days	
Incentive bonus			
Shipments			
New accounts generated			

Table 4.2. Examples of soft data.

WORK HABITS	FEELINGS/ ATTITUDES	NEW SKILLS	DEVELOPMENT/ ADVANCEMENT	WORK CLIMATE	INITIATIVE
Absentism	Favorable reactions	Decisions made	Number of promotions	Number of grievances	Implementation of new ideas
Tardiness	Attitude changes	Problems solved	Number of pay increases	Number of discrimination charges	Successful completion of projects
Visits to the dispensary	Perceptions of job responsibilities	Conflicts avoided	Number of training programs attended	Employee complaints	Number of suggestions submitted
First-aid treatments	Perceived changes in performance	Grievances resolved	Requests for transfer	Job satisfaction	Number of suggestions implemented
Violations of safe	Employee loyalty	Counseling problems solved	Performance appraisal ratings	Employee turnover	Work accomplishment
Number of communication breakdowns	Increased confidence	Listening skills	Increases in job effectiveness	Litigation	Setting goals and objectives
Excessive breaks		Interviewing skills			
Follow-up		Reading speed			
		Discrimination charges resolved			
		Intention to use new skills			
		Frequency of use of new skills			

evaluation decisions such as what data to collect, who the sources will be, when the data should be collected, how the effects of the program will be isolated, etc. The broad purposes of evaluation can be categorized into 10 evaluation purposes, identified and described below:

To determine success in accomplishing training objectives. Every training initiative should have objectives, stated in a generally accepted format (i.e., measurable, specific, challenging, etc.). Evaluation provides input to determine if objectives are being (or have been) met.

To identify the strengths and weaknesses in the training process. Probably the most common purpose of evaluation is to determine the effectiveness of the various elements and activities of the training process. Training components include, but are not limited to, methods of presentation, learning environment, training content, learning aids, schedule, and the facilitator. Each component makes a difference in the training effort and must be evaluated to make improvements in the training process.

To compare the costs to the benefits of a training program. With today's business focus on the bottom line, determining a training program's cost-effectiveness is crucial. This evaluation compares the cost of an training to its usefulness or value, measured in monetary benefits. The return on investment is the most common measure. This evaluation measure provides management with information needed to eliminate an unproductive intervention, increase support for training programs that yield a high payoff, or make adjustments in an program to increase benefits.

To decide who should participate in future programs. Sometimes evaluation provides information to help prospective participants decide if they should be involved in the program. This type of evaluation explores the application of the program to determine success and barriers to implementation. Communicating this information to other potential participants helps decide about participation.

To test the clarity and validity of tests, cases, and exercises. Evaluation sometimes provides a testing and validating instrument. Interactive activities, case studies, and tests used in the training process must be relevant. They must measure the skills, knowledge, and abilities the program is designed to teach.

To identify which participants were the most successful with the training. An evaluation may identify which participants excelled or failed at learning and implementing skills or knowledge from the training. This information can be helpful to determine if an individual should be promoted, transferred, moved up the career ladder, or given additional assignments. This type of evaluation yields information on the assessment of the individual or groups in addition to the effectiveness of the training.

To reinforce major points made to the participant. A follow-up evaluation can reinforce the information covered in a training program by attempting to measure the results achieved by participants. The evaluation process reminds participants what they should have applied on the job and the subsequent results that should be realized. This follow-up evaluation serves to reinforce to participants the actions they should be taking.

To gather data to assist in marketing future programs. In many situations, training departments are interested in knowing why participants attend a specific program, particularly where many programs are offered. An evaluation can provide information to develop the marketing strategy for future programs by determining why participants attended, who made the decision to attend, how participants found out about the training, and if participants would recommend it to others.

To determine if the training was the appropriate solution for the specific need. Sometimes evaluation can determine if the original problem needed a training solution. Too often, a training intervention is conducted to correct problems that cannot be corrected by training. There may be other reasons for performance deficiencies, such as procedures, work flow, or the quality of supervision. An evaluation may yield insight into whether or not the training intervention was necessary, and possibly even point management toward the source of the problem.

To establish a database that can assist management in making decisions. The central theme in many evaluations is to make a decision about the future of a training initiative. This information can be used in positions of responsibility, including instructors, training staff members, managers (who approve training), and executives (who allocate resources for future programs). A comprehensive evaluation system can build a database to help make these decisions.

APPROACHES TO COLLECTING BASELINE DATA

Baseline data can often be collected before a program begins by simply obtaining it from the records of the organization. For example, sales, quality, and output data are often available. Also, data such as turnaround time for producing a product and getting it to a customer, customer-satisfaction scores, employee-satisfaction scores, etc. are often available. But the issue of collecting baseline and outcome data hinges on more than availability. It also must match the group being trained.

For example, suppose that customer-satisfaction scores are kept by the organization as a unit, but not kept on each individual customer service rep. One hundred percent of the customer-service representatives from this business unit are being trained in customer-service skills. The customer satisfaction scores of the unit would represent a measure that is influenced by the group being trained. There also may be other influences, but we are at least certain that the entire group has the opportunity to influence the scores following a training intervention. Therefore, the scores of the business unit would be representative of the population being trained. However, if, for example, only half of the reps are being trained, then the scores of the entire business unit are not representative. We would need to know how the scores are influenced by each individual trained in order to have useful data.

When a scenario such as the one presented above prevents us from using data available in the organization, we must use alternative ways to collect both baseline and outcome data. For example, Figure 4.1 shows how a questionnaire can be structured to solicit before-and-after data from the participant or from managers in the business unit. This type of data is easily estimated if the respondent is made aware that this will be asked following the training and if the questionnaire is administered within a few months after the training.

When the client or a client representative must provide outcome measures and baseline data, it is often helpful to ask a series of questions such as those listed on the checklist in Table 4.3.

KEY FACTORS TO CONSIDER WHEN DEVELOPING AN EVALUATION STRATEGY

The data collection plan and ROI analysis plan make up the evaluation strategy to implement the evaluation project. As this strategy is developed, a number of factors must be considered to ensure that the strategy is practical and can be accomplished within the framework of the budget and other required resources. These factors are listed here.

1. Location of participants
2. Duration of program
3. The importance of program in meeting organizational objectives
4. Relative investment in the program
5. Ability of participants to be involved in evaluation
6. The level of management interest and involvement in the process

Figure 4.1. Downloadable questionnaire example soliciting before-and-after performance.

QUESTION	STRONGLY DISAGREE			STRONGLY AGREE		
	1	2	3	4	5	
I keep a good record of things that are done and said in meetings that I attend.						
						Before Training
						After Training
I prioritize my job tasks so that the most important aspects of my job get the most time and attention.						
						Before Training
						After Training
My communication at work is very effective.						
						Before Training
						After Training
I have a good sense of control over my job.						
						Before Training
						After Training
I record meetings and appointments on a monthly calendar.						
						Before Training
						After Training
I begin each day with a planning session.						
						Before Training
						After Training
I make a daily task list.						Before Training
						After Training
I am on time for my appointments and meetings.						Before Training
						After Training

Table 4.3. Downloadable checklist for outcome and baseline data.

OUTCOME DATA ASK THE CLIENT
✔ What do you (the client) want to change?
✔ What will it look like when it has changed?
✔ How does it look now?
✔ What direct measures reflect the change? (Output, Quality, Cost, Time)
✔ What indirect measures reflect the change?
✔ Who can provide information about the relationship between training and the business measures?
✔ What other factors will influence these business measures?
✔ What solutions have you tried?
BASELINE DATA ASK THE CLIENT
✔ Is baseline data available from organization records?
✔ Does the data available match the population being trained (can it be traced exactly to the trained population without contamination)?
✔ If organization does not have baseline data, can it be estimated by participants or others?
✔ What is the best timing for the baseline data collection?

7. The content and nature of the program
8. Senior management interest in evaluation
9. The availability of business results measures

These factors will play a major role in influencing data collection methods and other evaluation decisions. They can affect the feasibility of the project and therefore the purpose and level of the evaluation being pursued. It is best to consider these factors thoroughly during the planning phase of evaluation.

DEVELOPING EVALUATION PLANS AND STRATEGY

There are two types of evaluation plans that must be developed prior to implementing the evaluation initiative. The data collection plan is the initial planning document and it is already partially developed when the program objectives and measures are developed from the instructions above and from the information presented in the previous chapter. The ROI analysis plan follows the development of the data collection plan, and together they comprise the evaluation strategy.

Data Collection Plan. Table 4.4 illustrates the data collection plan *worksheet* and Table 4.5 illustrates an *example* using a sales training program as the program being evaluated.

Using Table 4.5 and columns A through F labeled as a reference, the following steps are involved in completing the data collection plan. The broad objectives (*column A*) have already been established using the techniques illustrated in Chapter 5. The measures (*column B*) are determined based on the capability to measure each objective at evaluation levels 1 through 5. *Column C* lists the data collection methods to be used for the evaluation project. Nine possible methods to collect data are listed below. Choices can be made from this list of possibilities. Choices are based on the constraints of disruption, the cost, the availability of the data, the quality of the data, and the willingness and cooperation of data sources.

Table 4.4. Downloadable data collection plan worksheet.

PROGRAM: _____ RESPONSIBILITY: _____
DATE: _____

LEVEL	BROAD PROGRAM OBJECTIVE(S)	MEASURES	DATA COLLECTION METHOD/INSTRUMENTS	DATA SOURCES	TIMING	RESPONSIBILITIES
1	Reaction/satisfaction and Planned Actions					
2	Learning					
3	Application/ implementation					
4	Business impact					
5	ROI					

Comments: _____
_____
_____

Table 4.5. Data collection plan example.

Program: _____   Responsibility: _____   Date: _____

Level	Column A Broad Program Objective(s)	Column B Measures	Column C Data Collection Method/Instruments	Column D Data Sources	Column E Timing	Column F Responsibilities
1	**REACTION/SATISFACTION** • Positive reaction • Recommended improvements • Action Items	• Average rating of at least 4.2 on 5.0 scale on quality, -usefulness, and achievement of program objectives. • 100% submission of planned actions	• Reaction Questionnaire	• Participants	• End of 2nd day • End of 3rd day	• Facilitator
2	**LEARNING** • Acquisition of Skills • Selection of Skills	• Through live role-play scenarios, demonstrate appropriate selection and use of all 15 sales interaction skills and all 6 customer influence steps.	• Skill Practice	• Participants	• During program	• Facilitator
3	**APPLICATION/IMPLEMENTATION** • Use of Skills • Frequency of Skill Use • Identify Barriers	• Reported frequency and effectiveness of skill application • Reported barriers to customer interaction and closing of sales.	• Questionnaire • Follow-up Session	• Participants	• 3 Months after Program • 3 Weeks After Program	• Training Coordinator • Facilitator
4	**BUSINESS IMPACT** • Sales increase	• Weekly sales	• Performance Monitoring	• Store records	• 3 months after Program	• Training Coordinator
5	**ROI** Target ROI at least 25%					

Comments: _____   *Get stakeholder buy-in to control group arrangement and the criteria to establish the control group.* _____

Return on Investment and Training Performance Improvement Programs, Jack J. Phillips, Ph.D. Butterworth-Heinemann, 1997.

1. Follow-up surveys measure satisfaction from stakeholders.
2. Follow-up questionnaires measure reaction and uncover specific application issues with training interventions.
3. On-the-job observation captures actual application and use.
4. Tests and assessments are used to measure the extent of learning (knowledge gained or skills enhanced).
5. Interviews measure reaction and determine the extent to which the training intervention has been implemented.
6. Focus groups determine the degree of application of the training in job situations.
7. Action plans show progress with implementation on the job and the impact obtained.
8. Performance contracts detail specific outcomes expected or obtained from the training.
9. Business performance monitoring shows improvement in various performance records and operational data.

Column D documents the data sources to be used. Possible sources are organizational performance records, participants, supervisors of participants, subordinates of participants, management, team/peer group, and internal/external groups such as experts.

Column E is used to document the timing of the evaluation. Timing is often dependent upon the availability of data (when is it available and does this time frame meet my needs?), the ideal time for evaluation (when will the participants have the opportunity to apply what they learned?), the ideal time for business impact (how long after application at level 3 will the impact occur at level 4?), convenience of collection, and constraints of collection (such as seasonal influences in the measurement being studied).

Column F lists the responsibilities for data collection. Level 1 and 2 are usually the responsibility of the facilitator/instructor or program coordinator. Level 3 and 4 data should be collected by a person having no stakes in the outcome. This avoids the appearance of bias that can be an issue to the credibility of the evaluation project.

Next, the ROI Analysis Plan must be developed. This plan determines the methods that will be used to analyze the data. This includes isolating the effects of the training, converting data to monetary values, capturing the costs of the training, and identifying intangible benefits. Table 4.6 is the ROI Analysis Plan *Worksheet* and it is illustrated in an *example* in Table 4.7 using the same sales-training program.

Using Table 4.7 and columns A through H labeled as a reference, the following steps are involved in completing the ROI Analysis Plan. *Column A* is used to plot the level-4 business impact measures from the *example* data collection plan, Table 4.5. Weekly sales per employee is a detailed description of the measure at level 4. When there is more than one level-4 measure, that measure is listed under the previous measure in the space labeled row B, and so on for each successive measure. Each level-4 measure must be addressed independently to determine and document the method of isolation and the method of conversion. What works well for one measure may not work so well for another. This example plan only has one level-4 measure. The methods to isolate the effects (column B) are listed below. At least one of these methods must be selected to isolate the effects. The methods for isolating the effects are:

1. A pilot group receiving the training is compared to a control group that does not receive the training to isolate the program's impact.
2. Trend lines are used to project the values of specific output variables, and projections are compared to the actual data after a training program is implemented.
3. A forecasting model is used to isolate the effects of a training program when mathematical relationships between input and output variables are known.
4. Participants/stakeholders estimate the amount of improvement related to training.

Table 4.6. Downloadble ROI analysis plan.

PROGRAM: _____   RESPONSIBILITY: _____
DATE: _____

DATA ITEMS (USUALLY (LEVEL 4)	METHODS FOR ISOLATING THE EFFECTS OF THE PROGRAM/ PROCESS	METHODS OF CONVERTING DATA TO MONETARY VALUES	COST CATEGORIES	INTANGIBLE BENEFITS	COMMUNICATION TARGETS FOR FINAL REPORT	OTHER INFLUENCES? ISSUES DURING APPLICATION	COMMENTS

Table 4.7. Example: ROI analysis plan.

PROGRAM: _Interactive Selling Skills_   RESPONSIBILITY: _____   DATE: _____

Column A	Column B	Column C	Column D	Column E	Column F	Column G	Column H
DATA ITEMS (USUALLY (LEVEL 4)	METHODS FOR ISOLATING THE EFFECTS OF THE PROGRAM/ PROCESS	METHODS OF CONVERTING DATA TO MONETARY VALUES	COST CATEGORIES	INTANGIBLE BENEFITS	COMMUNI-CATION TARGETS FOR FINAL REPORT	OTHER INFLUENCES? ISSUES DURING APPLICATION	COMMENTS
Weekly sales per employee	_Control group analysis_	_Direct conversion conversion using profit contribution_	_Facilitation fees Program materials_ _Meals/ refreshments_ _Facilities_ _Participant salaries/benefits_ _Cost of coordination/ evaluation_	_Customer satisfaction_ _Employee satisfaction_	_Program participants_ _Electronics dept. Manager-target stores_ _Store managers-target stores_ _Senior store executives district, region, headquarters_ _Training staff: instructors, coordinators, designers, and managers_	_No communication with control group_	_Must have job coverage during training_

Return on Investment and Training Improvement Programs, Jack J. Phillips, Ph.D. Butterworth-Heinemann, 1997.

5. Supervisors estimate the impact of a training program on the output measures.

6. Management can sometimes be used as a source to estimate the impact of a training program when numerous factors are influencing the same variable and a broader view is necessary to determine an estimate.

7. External studies provide input on the impact of a training program.

8. Independent experts provide estimates of the impact of a training program on the performance variable.

9. When feasible, other influencing factors are identified and the impact is estimated or calculated, leaving the remaining, unexplained improvement attributable to the training program.

10. Customers provide input on the extent to which the skills or knowledge of an employee have influenced their decisions to use a product or service.

It is possible to have very good data at level 4, but the ability to convert it to a monetary value does not exist, or excessive resources may be required to make the conversion. When this is the case, the data falls into the intangible category because we do not know what the benefits from the solution/training are worth. This is still important and useful data, but we are left wondering if the training solution costs more than the resulting benefits. So, whenever possible and practical, we want to convert the benefits into a monetary value as long as our conversion is credible. We must ask the question: can I easily convince others that the value, even though it may be an estimate, is a worthwhile expression of the contribution? If the answer is yes, then proceed. If the answer is no, consider being satisfied with level 4, intangible data. The methods used to convert data to a monetary value (*column C*) are identified below;

1. Output data are converted to profit contribution or cost savings and reported as a standard value.

2. The cost of a quality measure, such as a reject or waste, is calculated and reported as a standard value.
3. Employee time saved is converted to wages and benefits.
4. Historical costs of decreasing the negative movement of a measure, such as a customer complaint, are used when they are available.
5. Internal or external experts estimate a value of a measure.
6. External databases contain an approximate value or cost of a data item.
7. Participants estimate the cost or value of the data item.
8. Supervisors or managers provide estimates of costs or value when they are both willing and capable of assigning values.
9. The training staff estimates the value of a data item.
10. The measure is linked to other measures for which the costs are easily developed.

Column C is used to determine and document the method to be used to convert each data item to a monetary value. Once this decision is made, the key components are in place to analyze the benefits from the training's impact. The remaining columns on the ROI Analysis Plan are developed with the entire evaluation project in mind. *Column D* is used to determine the cost categories that will be targeted to collect the costs of the training program being evaluated.

Column E is used to determine and document the expected intangible benefits. This could be data that should be influenced by the program and is known to be difficult to convert to a monetary value, or it could be data that the stakeholders are not interested in converting.

Column F should list the stakeholders that the communication (results from the study) will target. By determining who these stakeholders are, two goals can be met. First, their ideas, opinions, and expectations about the study can be solicited and managed and the project can be designed with this in mind. Second, the proper reports can be developed to suit each audience.

Column G can be used to document any issues or influences that will impact the thoroughness, timeliness, objectivity, or credibility of the evaluation project. *Column H* is simply for comments.

FURTHER READING

Kaufman, Roger, Sivasailam Thiagarajan, and Paula MacGillis, editors. *The Guidebook for Performance Improvement: Working with Individuals and Organizations.* San Francisco: Jossey-Bass/Pfeiffer, 1997.

Kirkpatrick, Donald L. *Evaluating Training Programs: The Four Levels*, 2nd Edition. San Francisco: Berrett-Koehler Publishers, 1998.

Phillips, Jack J. *Handbook of Training Evaluation and Measurement Methods*, 3rd Edition. Houston: Gulf Publishing, 1997.

Phillips, Jack J. 1994, 1997. *Measuring The Return On Investment.* Vol 1 and Vol 2. Alexandria, VA: American Society for Training and Development.

Phillips, Jack J. "Return On Investment in Training and Performance Improvement Programs." Houston, TX: 1997, Gulf Publishing.

Phillips, Jack J. "Was It The Training?" *Training & Development*, Vol. 50, No. 3, March 1996, pp. 28-32.

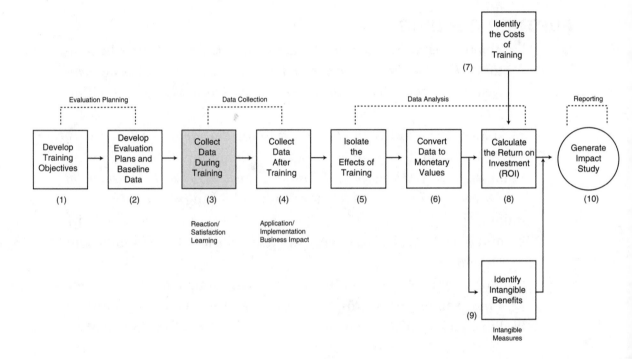

5

Step 3. Collect Data During Training (Levels 1 and 2)

Data collection can be considered the most crucial step of the evaluation process because without data, there can be no evidence of program impact and therefore no evaluation study. Two chapters are devoted to data collection. This chapter focuses on collecting data during implementation to determine participant reaction and/or satisfaction (level 1) and learning (level 2). Chapter 6 focuses on follow-up evaluation to determine application and/or implementation (level 3) and business impact (level 4).

It is necessary to collect data at all four levels because of the chain of impact that must exist for a training program to be successful. To recap the chain of impact: participants in the program should experience a positive reaction to the program and its potential application and they should acquire new knowledge or skills to implement as a result of the program. As application/implementation opportunities are presented, there should be changes in on-the-job behavior that result in a positive impact on the organization. The only way to know if the chain of impact has occurred is to collect data at all four levels.

LEVEL 1: MEASURING REACTION AND SATISFACTION

The importance of level-1 measurement

This chapter presents some typical measurement forms for level 1 and some supplemental possibilities. A level-1 evaluation collects reactions to the training program and can also indicate specific action plans. Measurement and evaluation at level 1 is important to ensure that participants are pleased with the training and see its potential for application on the job. However, level-1 results are not appropriate indicators of trainer performance; level-2 and level-3 results are much more appropriate indicators of this.

In many organizations, 100% of training initiatives are targeted for level-1 evaluation. This is easily achievable because it is simple, inexpensive, and takes little time to design the instruments and to collect and analyze the data.

LEVEL OF DATA	TYPE OF DATA
1	Satisfaction/reaction, planned action
2	Learning
3	Application/implementation
4	Business impact
5	ROI

Because a level-1 evaluation asks participants only for their reactions to the training, making decisions based only on level-1 data can be unwise. However, such data can be useful in identifying trends, pinpointing problems in program design, and making improvements in program delivery and timing. It can also be useful in helping participants to think about how they will use what they learned when they return to the work setting. Even if a level-3 evaluation is planned for later, information can be collected with a level-1 instrument to identify barriers to the transfer of training to the job. Beyond these applications, level-1 data is of little use. It is also helpful to understand that level-1 data is often inflated because participants don't always give candid responses regarding the training and, even when they do, they may be influenced by the recent attention they have received from the trainer.

Methods of level-1 data collection

Two basic methods are used to collect level-1 data: questionnaires and interviews. The most common and preferred method is a questionnaire. Several different types of items can be used on a questionnaire. These include:

Sample response items for a level 1 questionnaire.	
TYPE	**EXAMPLE**
Scaled rating	Strongly Disagree — 1, Disagree — 2, Neutral — 3, Agree — 4, Strongly Agree — 5
Open-end items	Which part of the training did you like best?
Multiple choice	Which of the following items might keep you from using the skills you learned? (Check all that apply)
	❏ My supervisor doesn't agree with the procedures taught.
	❏ The people I work with don't know how to do this.
	❏ If I do it, I won't meet my performance objectives.
Comments	Overall comments:
Comparative rankings	Rank order the following parts of the course for their usefulness to you on your job, with 1 representing the most useful and 10 representing the least useful.

The second method is the interview. This may be conducted in a one-on-one setting, in a focus group, or over the telephone. An interview guide should be developed that includes written questions/items similar to those used in questionnaires. This will help ensure consistency and completeness. In most instances, a questionnaire is used to obtain level-1 data because of the time and costs required to conduct interviews.

This chapter presents information about using questionnaires. The interview process is described in more detail in Chapter 6, which deals with level-3 evaluation.

LEVEL-1 TARGET AREAS—STANDARD FORM

Figure 5.1 shows a typical form for level-1 evaluation. It should be customized to fit the organization and specific training objectives. The major sections are as follows:

- Section I collects information on content issues, including success in meeting program objectives.
- Section II examines the training methodology and key issues surrounding the training materials used in the program.
- Section III covers the learning environment and course administration—two very key issues.
- Section IV focuses directly on the skills and success of the trainer. Input is solicited for areas such as knowledge base, presentation skills, management of group dynamics, and responsiveness to participants. [Because level-1 participant feedback is easily influenced, it should never be used as the sole input regarding the performance of the trainer. Greater value should be placed on level-2 (knowledge and skills learned) achievements. It is also helpful to compare level-2 achievement with level-1 reaction feedback to determine if there is a correlation between high level-2 scores and high level-1 scores.]
- Section V, an important part of the evaluation, provides ample space for participants to list planned actions directly linked to the training.
- Section VI completes the evaluation with an overall program rating, on a scale of poor to excellent.

Figure 5.1. Typical level-1 evaluation form to be completed and returned at the conclusion of this program.

Typical Level-1 Evaluation Form
To Be Completed and Returned at the
Conclusion of This Program

Training Initiative Name:_____ Date: _____

Training Initiative Number:_____ Location: _____

Training and Development values your comments.

The statements below concern specific aspects of this program. Please indicate to what extent you agree or disagree with each statement and **provide your comments** where appropriate, using the following scale:

❶	❷	❸	❹	❺	❻
Strongly Disagree	Disagree	Neutral	Agree	Strongly Agree	Not Applicable

	❶	❷	❸	❹	❺	❻
I. Content						
1. Objectives were clearly explained	O	O	O	O	O	O
2. Objectives stated were met	O	O	O	O	O	O
3. I understand the materials and topics in this program	O	O	O	O	O	O
4. Content is relevant to my job (if not, please explain)	O	O	O	O	O	O

Your comments, please: (Please print)

	❶	❷	❸	❹	❺	❻
II. Methodology - The following activities/materials helped me to understand the content and achieve the stated objectives.						
5. Pre-work received prior to program	O	O	O	O	O	O
6. Participant's workbook	O	O	O	O	O	O
7. Class discussions	O	O	O	O	O	O
8. Exercises and/or readings/activities	O	O	O	O	O	O
9. Audio/Visuals (flip charts, videos, etc.)	O	O	O	O	O	O

Your comments, please: (Please print)

	❶	❷	❸	❹	❺	❻
III. Environment/Administration						
10. The classroom was suitable for this type of program	O	O	O	O	O	O

Your comments, please: (Please print)

Figure 5.1. (*continued*).

Name of Facilitators	#1 _____						#2 _____						#3 _____					
	❶	❷	❸	❹	❺	❻	❶	❷	❸	❹	❺	❻	❶	❷	❸	❹	❺	❻
IV. Facilitator																		
11. Appeared knowledgeable of the subject matter	○	○	○	○	○	○	○	○	○	○	○	○	○	○	○	○	○	○
12. Presented clearly to assist my understanding	○	○	○	○	○	○	○	○	○	○	○	○	○	○	○	○	○	○
13. Promoted discussion and involvement	○	○	○	○	○	○	○	○	○	○	○	○	○	○	○	○	○	○
14. Responded appropriately to questions	○	○	○	○	○	○	○	○	○	○	○	○	○	○	○	○	○	○
15. Effectively managed group dynamics	○	○	○	○	○	○	○	○	○	○	○	○	○	○	○	○	○	○
16. Kept the discussion/ activities focused on stated objectives	○	○	○	○	○	○	○	○	○	○	○	○	○	○	○	○	○	○

Your comments, please: (Please print)

V. Planned Actions

17. As a result of this program, what will you do differently?

VI. Overall Program Rating

1 = Completely Unacceptable . . . . 10 = Very Exceptional	❶	❷	❸	❹	❺	❻	❼	❽	❾	❿
18. My overall rating for this program	○	○	○	○	○	○	○	○	○	○

Your comments, please: (Please print)

What may keep you from applying what you have learned in this program?
_____

Which target group is best suited for this program? Could you recommend specific individuals to attend?
_____

Please share any information you believe would help us to improve this program.

Thank you for taking the time to share your comments and reactions to your learning experience,.

- The form is complete with two additional open-ended questions concerning improvements and marketing information.

Enhancing level-1 evaluation

Impact and ROI. In some situations, additional questions related to the effects of planned actions may be sought on a level-1 reaction questionnaire as a supplemental document. Figure 5.2 shows the type of questions that may be asked on a level-1 questionnaire to collect information about how participants plan to apply what they have learned, in the form of estimated impact and ROI data. While this is a projection and an estimate, the data may be useful in several ways.

Figure 5.2. Estimated impact and ROI questions to supplement level-1 feedback questionnaires.

Projected Impact of Training

As a result of this program what do you estimate to be the increase in your personal effectiveness, expressed as a percentage?
_____%

Please indicate (specifically) what you will do differently on the job as a result of this program.

1. _____

2. _____

3. _____

As a result of any changes in your thinking, new knowledge, or planned actions, please estimate (in monetary values) the benefits to your organization (e.g., reduced absenteeism, reduced employee complaints, better teamwork, increased personal effectiveness) over a period of one year.

$ _____

What is the basis of this estimate? _____

_____

What confidence, expressed as a percentage, can you put in your estimate? (0%=No Confidence; 100%=Certainty) _____%

These questions require participants to think beyond the mere application of training and to consider business impact and monetary values. Analysis of this data yields a projected ROI. Of course, there is no guarantee that these results will materialize. The following adjustments should be made to make the forecast more realistic and conservative.

- The total benefits are based only on the improvement data furnished by the participants. The participants who do not furnish data or who provide incomplete data are assumed to have no planned improvement. If a participant is unwilling or unable to provide information about how he or she will apply the skills learned, it is inappropriate to infer that the person will have an improvement.

- The costs should be fully loaded and include values for all participants.
    - Include participants' estimated salaries and benefits for the time that the participants are in training on company time. The Human Resources department can usually provide current salary averages and a percentage benefit factor to use in calculations.
    - Include an opportunity cost over and above salaries and benefits when it is appropriate to do so. An example of opportunity cost is lost sales or lost production while participants are in training.
    - If sales reps or other commissioned employees are paid while in training, this amount should be included in calculations.
    - Facility costs, food and refreshments, and AV equipment rental should be included.
    - Costs of research, needs analysis, and evaluation should be allocated to the program being evaluated.
    - All delivery costs, direct and indirect, should be included.
    - Overhead from the training and development should be allocated.

- The improvement values should be adjusted to reflect the confidence level. Each value is multiplied by the respective confidence percentage. For example, if a participant predicts a $20,000 improvement with an 80% confidence level, the participant is suggesting a potential 20% error in the estimation. The value could be in the range of $16,000 to $24,000. The $16,000 value is used to be conservative.
- Only the first-year values for benefits should be used. If successful, the program should produce improvements for several years.
- When calculating a level 1 projection, a "drop-off" factor is used to reduce the estimated value. Experience has shown that the results predicted at the end of a program do not fully materialize. Unless an organization has developed a drop-off factor based on experience, a factor of $1/2$ is used. In essence, only half of the projection is used in the analysis improvement.

While this process is subjective and sometimes unreliable, it does have some usefulness.

- If evaluation must stop at this level, this approach provides more insight into the value of the training initiative than the data from typical reaction questionnaires.
- Managers and executives will usually find this data more useful than a report stating, for example, that "40% of participants rated the training above average."
- This data can form a basis for comparison of different presentations of the same training. If one training program results in a forecast ROI of 300% and another results in a forecast of 30%, it appears that one presentation may be more effective than the other.
- Collecting this type of data focuses increased attention on program outcomes. Participants leave the program with an understanding that specific behavioral change is expected. Even if the projected improvement data are discarded, the supplemental evaluation is productive because of the important message sent to participants and the mental exercises of estimating impact and monetary value.

- If a follow-up is planned to determine impact, the data collected in the level-1 projection can be very helpful for comparison. Level-1 ROI data collection helps participants plan the implementation of what they have learned. (Incidentally, when participants know that a follow-up is planned, they are more conservative with these estimates.)

Other Improvement Issues. In addition to the standard items, a level-1 evaluation of a training program may have specific information needs. For example, a training program may have different applications for different groups that would require specific questions from those groups. For example, a question may be developed specifically for sales representatives. A similar question may be designed for the information system or marketing staff. In practice, the standard form will suffice for the majority of evaluations. A supplemental form should be used only when it is necessary to develop additional information; then the following issues should be explored when developing the additional questions or questionnaires.

- The reason for the additional evaluation. What decisions will be made as a result of the findings?
- Who will make decisions based on the evaluation results? Who is the target audience of the evaluation report?

Keep the additional questionnaire simple. As a general rule, it should be no more than one 2-sided page and should take no longer than 10–15 minutes to complete.

Administrative issues

Several administrative issues are critical to the success of a level-1 evaluation. The following guidelines should be used when appropriate.

- The name of a participant should not be identified unless it has been volunteered or is necessary to link with other data.
- Participants should be told how the data will be used (e.g., to improve the training program or certain elements of the training) and who will see the results (e.g., provided in summary form to management groups).

- Participants should be encouraged to provide complete and thorough information. The importance of their feedback should be emphasized.
- The best time to administer a level-1 evaluation is at the completion of the training program when materials, skills, and topics can be recalled easily.
- If possible, the training coordinator or sponsor should collect the evaluation questionnaires, not the actual trainer. This ensures that candid information is provided and helps to keep the evaluation process objective.
- For longer training programs, a specified time period may be allocated to complete the questionnaire even before the end of the training program. This will allow ample time for providing adequate quality and quantity of information.
- If diagnostic-type data on a particular part of the training program is desired, an evaluation regarding that part of the training program should be administered immediately after it is completed.
- For lengthy training programs (more than one day), an end-of-day questionnaire may be appropriate. The questionnaire can be a daily feedback form or a standard questionnaire designed for specific parts of the training program. This approach allows for evaluation of important topics while the material can be recalled easily. It also allows the trainer to make daily adjustments. Figure 5.3 shows a form that can be used to collect daily feedback.
- For the typical training program where there is one trainer, a *checklist worksheet* to assist in the development of an end-of-program questionnaire is shown in Figure 5.4 as an example of this type of a level-1 questionnaire.
- For program sessions in which multiple speakers and/or topics are presented, a form should be used that can capture data in a concise manner (Figure 5.5). Figure 5.5 shows an example that can be used to create your own questionnaire.

Figure 5.3. Downloadable form.

Daily Feedback Form
1. What issues presented today still remain confusing and/or unclear?
2. Which topics are most useful?
3. It would help me if you would
4. The pacing of the program is:
❑ Just Right
❑ Too Slow
❑ Too Fast
5. The degree of involvement of participants is:
❑ Not Enough
❑ Too Much
❑ Just Right
6. The three items that are very important for me that you should cover tomorrow are:
7. Comments

Figure 5.4. Downloadable checklist.

Checklist	
Consider asking about	
1. Clarity of program objectives	✔
2. If what was learned will be useful in participant's work	✔
3. If the order of the program topics and activities was logical	✔
4. If the pace of the program was good—neither too fast nor too slow	✔
5. If the program materials were easy to use	✔
6. If the program materials were relevant	✔
7. If exercises and job simulations were relevant	✔
8. If participants would recommend the program to others	✔
9. If the program is a good investment for the organization	✔
10. Participant's confidence level to do the job before and after training	✔
11. If skill practice was sufficient	✔
12. If trainer was knowledgeable	✔
13. If trainer was well organized	✔
14. If trainer adequately handled participant's questions	✔
15. If trainer kept participants actively involved	✔
16. What specific actions participant will do differently on returning to the job	✔
17. The key performance area(s) for the savings or contribution from actions taken	✔
18. What specific measures or outcomes will change as a result of actions taken on the job	✔
19. Estimating (in monetary values) the benefits to the organization	✔
20. Confidence level participant places on the monetary estimate	✔

LEVEL 2: MEASURING LEARNING

Level-2 measurement is important in evaluating the success of training programs in terms of whether the participants acquired the desired knowledge or skill. A variety of techniques can be used to measure learning in addition to the traditional approach of using objective tests. The use of these techniques is often constrained by budgets, perceptions, time limitations, and the ability to synthesize learning and skill components or simulate actual settings in which skills are applied.

LEVEL OF DATA	TYPE OF DATA
1	Satisfaction/reaction, planned action
2	Learning
3	Application/implementation
4	Business impact
5	ROI

A level-2 evaluation is simply a measure of whether or not a participant can demonstrate the level of knowledge and/or skill required by the objectives. This evaluation usually occurs during or at the end of a training program and can be accomplished in a variety of ways, as is illustrated later in this chapter.

The need to measure learning

There are four key areas that demonstrate why learning is an important measure in evaluating a training program. Any of these, individually, would probably justify the need to measure learning. Collectively, they provide a major thrust for measuring the amount of knowledge, skills, or behavioral change that results from a training program.

The need for transfer of learning. A significant problem that has plagued the training and development field for many years is a lack

Figure 5.5. Downloadable participant's opinion form.

PROGRAM TITLE: _____ LOCATION: _____
TRAINER: _____ BEGINNING DATE: _____

PART I: DEMOGRAPHIC INFORMATION

Check the appropriate answer to each question.

	a.	b.	c.	d.	e.
1. How long have you worked for the company?	2 years or less	3-5 years	6-10 years	11-20 years	20+ years
2. How long have you held your current job?	less than one year	1-2 years	3-5 years	6-10 years	over 10 years
3. In which area do you work?	a. Corporate	b. Customer Svc/Acct. Support	c. Marketing	d. Sales	e. Other

Part II: Your Reaction To This Program

Circle one rating number for each item.

	Strongly Disagree	Disagree	Agree	Strongly Agree	Not Applicable
4. The program objectives were clear.	1	2	3	4	n/a
5. Overall, what I learned in this program will useful in my work.	1	2	3	4	n/a
6. The order of the program topics and activities made sense to me.	1	2	3	4	n/a
7. The pace of this program was good—neither too fast nor too slow.	1	2	3	4	n/a
8. The program materials were easy to use.	1	2	3	4	n/a
9. I am satisfied with what I gained from this program.	1	2	3	4	n/a
10. I would recommend this program to others.	1	2	3	4	n/a
11. This program is a good investment for the company.	1	2	3	4	n/a
12. Examples and illustrations helped me understand the material.	1	2	3	4	n/a

1. The overall rating I would give this program is:

not useful	useful	very useful	extremely useful
☐	☐	☐	☐

Figure 5.5. Participant's opinion form (*continued*).

PART III

PLEASE RATE THE PRESENTERS REGARDING THEIR OVERALL PRESENTATIONS, INCLUDING KNOWLEDGE OF THE SUBJECT, ABILITY TO CONVEY THE INFORMATION IN AN INTERESTING AND INFORMATIVE WAY, AND THE OVERALL VALUE OF THE PRESENTATION TO MEET YOUR LEARNING NEEDS. THE RATING SCALE IS SHOWN BELOW. PLEASE ADD ANY ADDITIONAL COMMENTS ABOUT THE BEST PART OF EACH PRESENTATION AND THE PART THAT COULD IMPROVE.

NAME	SUBJECT	RATING			
		Strongly Disagree	Disagree	Agree	Strongly Agree
Bill Best Part:_____ Could Improve:_____	*Financial Outlook* Knowledge of subject Ability to inform and hold interest Overall value of presentation	❏ ❏ ❏	❏ ❏ ❏	❏ ❏ ❏	❏ ❏ ❏
		Strongly Disagree	Disagree	Agree	Strongly Agree
Sue Best Part:_____ Could Improve:_____	*Goals Status* Knowledge of subject Ability to inform and hold interest Overall value of presentation	❏ ❏ ❏	❏ ❏ ❏	❏ ❏ ❏	❏ ❏ ❏
		Strongly Disagree	Disagree	Agree	Strongly Agree
Jose Best Part:_____ Could Improve:_____	*Status of New Product Development* Knowledge of subject Ability to inform and hold interest Overall value of presentation	❏ ❏ ❏	❏ ❏ ❏	❏ ❏ ❏	❏ ❏ ❏
		Strongly Disagree	Disagree	Agree	Strongly Agree
Amanda Best Part:_____ Could Improve:_____	*Supplier Relationships* Knowledge of subject Ability to inform and hold interest Overall value of presentation	❏ ❏ ❏	❏ ❏ ❏	❏ ❏ ❏	❏ ❏ ❏
		Strongly Disagree	Disagree	Agree	Strongly Agree
Bruce Best Part:_____ Could Improve:_____	*Strategic Direction* Knowledge of subject Ability to inform and hold interest Overall value of presentation	❏ ❏ ❏	❏ ❏ ❏	❏ ❏ ❏	❏ ❏ ❏
		Strongly Disagree	Disagree	Agree	Strongly Agree
Tuan Best Part:_____ Could Improve:_____	*The Competition* Knowledge of subject Ability to inform and hold interest Overall value of presentation	❏ ❏ ❏	❏ ❏ ❏	❏ ❏ ❏	❏ ❏ ❏
		Strongly Disagree	Disagree	Agree	Strongly Agree
Marie Best Part:_____ Could Improve:_____	*Customers and Marketing* Knowledge of subject Ability to inform and hold interest Overall value of presentation	❏ ❏ ❏	❏ ❏ ❏	❏ ❏ ❏	❏ ❏ ❏

of transfer of what is learned by participants from the training setting to the job setting. It is critical to make sure that learning occurs and that it is transferred to the job. In many situations, it is not. The result of the transfer is measured on the job during level-3 evaluation, as application/implementation. However, the actual learning that takes place needs to be measured early to see how much transfer can be expected and, in some cases, enhanced.

Increased emphasis on knowledge, expertise, and competence. Many organizations are focusing more on knowledge, expertise, and competence. Many large training initiatives involve developing expertise, with employees using tools and techniques not previously used. Other programs focus on core competencies and on building important knowledge, skills, and behaviors into the organization. With an increased focus on knowledge management, it is important for knowledge-based employees to acquire a vast array of information and use it in a productive way. This emphasis on employee knowledge and skills makes measuring learning in a training program crucial.

The importance of learning in a major change initiative. Although many change initiatives involve new processes, procedures, and technologies, the human factor is still crucial. Whatever the change, employees must learn how to work in the new environment and develop new knowledge, skills, and behaviors. Learning is becoming a larger part of change initiatives than previously because of the variety of tools, techniques, processes, and technologies involved. These must be used in an intelligent way to reap the benefits of the change initiative. Employees learn in a variety of ways, not just in a formal classroom environment, but also through technology-based learning and on-the-job facilitation with job aids and other tools. Team leaders, managers, and others are called on to reinforce, coach, and mentor in order to ensure that learning is transferred to the job and is implemented as planned.

Diagnosing what went wrong when there is a problem. When follow-up application and implementation does not go smoothly, the most important issue is to find out what went wrong and what needs to be adjusted. When learning is measured during the training implementation, it is easy to see the degree to which the lack of learning is or is not a problem. Without the learning measurement, the evaluator may not know why employees are not performing the way they should.

These key issues make learning an important aspect of training. It requires appropriate attention, as do the other five measures in the ROI process.

Issues in measuring learning

Several issues are involved in measuring learning from training programs. Each issue described below should be considered as the measurement process is developed.

Time and Cost for Development. Most level-2 measurement processes will take additional time to develop, test, and implement, adding to the overall cost of program development and administration. This extra time and cost must be considered when selecting the measurement technique, as well as deciding if a level-2 measure is actually required.

Time for Administration. A level-2 evaluation is usually conducted during the training program and will take precious time away from teaching or discussions related to program content. For some training programs, the duration has been mandated, and carving out additional time to measure learning becomes a critical issue. This requires learning measurement techniques to be efficient and brief.

Validity. A valid measurement process is necessary. A valid process is one that measures what it is designed to measure. While there are several approaches to measure validity, the most frequently used approach for training programs is the content-validity approach. In this situation, subject matter experts (individuals most knowledgeable with training program content) are asked to review the measurement process and to provide opinions as to whether the process actually measures what is necessary or needed on the job for the particular topic. This simple approach should suffice in most situations.

If the outcome of the measurement process leads to a specific human resource action for the individual, then other approaches to check for validity may be necessary. For example, if a participant is promoted, receives a pay increase, is allowed to enter a career development path, or is withheld from any of these actions because of performance during the measurement process, the validity of the process may be under greater scrutiny and potential challenge. A more precise and defensible approach to check validity may be necessary. Other references should be sought for this purpose.[1]

Reliability. The reliability of a measurement process refers to its consistency and stability over time. Ideally, the same measurement process used with the same participant in another time frame should provide the same performance, if the skills or knowledge have not been enhanced between those two time periods. The same comments about validity offered above apply to reliability. Reliability is more of an issue when the performance during the training program translates into a human resource action for the participant. Otherwise, the reliability can be enhanced by ensuring that all of the topics are covered in the measurement process and that the instructions are very clear and there is little room for different interpretations in responses. If a reliability check is necessary, several methods are available and are described in other resources.[1]

Success Determination. An important issue in the design of level-2 measurement is to determine the level of performance considered to be satisfactory. For an objective test, this is usually referred to as a cutoff score. For some measurement processes, such as measuring skill acquisition or understanding of a particular topic, the measure may be imprecise and the determination of a successful level of knowledge or skills may be difficult. In this situation, subject matter experts can be helpful in determining the level of performance that would be considered acceptable on the job. This should be determined in advance and, in some cases, communicated directly to participants, particularly if the measurement process is reduced to a specific score and they have knowledge of that score.

Confronting Failure. A final issue is the implication of an unsatisfactory learning outcome in the level-2 measurement process. In most situations this is not a serious issue. A participant may have an adverse action because of the deficiency. In these cases, several approaches may be considered, such as requiring additional study, repeating a portion of the material, suggesting an additional exercise to strengthen an area of weakness, or simply notifying the participant that there is a continued developmental need. The important point is that consequences need to be clearly defined so that they can be integrated into the administrative procedures.

Criteria to determine the need to measure learning

Not every training program will need to measure learning. Very short training programs (less than two hours) do not usually have a mechanism to measure learning except for what is captured with the level-1 evaluation form. In some other situation, it may not be necessary to measure the knowledge and/or skills acquired. The nature of the training program, the importance of the training, and the time and resources necessary to develop a level-2 evaluation may determine if level-2 measurement is necessary.

Some training programs will require level-2 evaluation. *Typical of these are:*

Training programs representing critical skills or knowledge that must be transferred to the job (e.g., because of regulations, profitability, or customer service needs). In these cases, it may be essential for participants to demonstrate that they have acquired the necessary knowledge and/or skills.

Training programs offered as part of a certification or licensing process, where the acquisition of knowledge or skills is essential.

Training programs where the program sponsor or client may be interested in assessing the extent of learning.

Training programs where a higher level evaluation is planned (levels 3, 4, and/or 5). In this situation, a level-2 evaluation is necessary; however, a less-comprehensive process may be permitted under these circumstances, since it is only necessary to have some evidence that learning has occurred.

Table 5.1 can be used to help determine whether level-2 measurement is necessary and, if so, to what degree.

Development of measures

If well-stated objectives have been developed, measures are easier to create. Here are a few suggestions:

- The performance measured in the training program should match the performance stated in the objectives.
- The performance conditions stated in the objective should also match the conditions under which the participant performs in the measurement process.
- If a measure that matches the objective cannot be developed, the objective should be reviewed. It may need to be revised or clarified.

If the course was not developed around well-stated objectives, the task may be more difficult. When this is the situation, here are a few suggestions:

- Instructions for individual and group activities such as exercises and case analysis should be reviewed. What are participants being asked to do in the activities? What's their purpose? Are they necessary? Can an objective be written from them?
- Materials such as a leader's guide, printed handouts and videotapes should be examined. Can objectives be written from them?

If objectives cannot be written from the review of all the course materials and activities, the training program may need revision.

Techniques for measuring learning

Measuring Learning with Formal Tests. Testing is important for measuring learning in training program evaluations. Pre- and post-

Table 5.1. (Downloadable form.) Determining the need
to measure learning.

SITUATION	TESTING CONSIDERATION	CHECK IF APPLICABLE
A) Learning objectives are not clearly stated	Create/redesign learning objectives before considering the need for testing learning	_____
B) Learning objectives indicate there is no need for mastery	You should test only to the level required by the learning objectives. Additionally, be certain that the sponsor and line managers are aware that mastery is not a requirement of the program and additional training may be required	_____
C) Learning objectives require skill application	Implement performance-based testing	_____
D) Program sponsor or client may be interested in assessing the extent of learning	Determine what type of evidence is acceptable and implement the appropriate method	_____
E) Interest lies in whether or not participants meet the desired minimum standards as spelled out in the learning objectives	Implement criterion-referenced testing	_____
F) Certification or licensing process is associated with the training program	Implement criterion-referenced and performance testing methods	_____
G) It is necessary to reproduce the job setting in a manner that is almost identical to the real setting; or safety is of paramount consideration	Use simulation techniques and be sure the budget will meet your needs	_____

Table 5.1. (Downloadable form.) Determining the need
to measure learning (*continued*).

SITUATION	TESTING CONSIDERATION	CHECK IF APPLICABLE
H) Critical skills or knowledge must be transferred to the job because of regulations, profitability, or customer service requirements	Implement performance testing methods	_____
I) A higher level of evaluation is planned	Consider the least-expensive testing methods since evaluations at lower levels do not need to be comprehensive under these circumstances	_____
J) Test results are used as criteria to determine promotions or career movement	Use objective testing methods (paper and performance) that have met strict validity and reliability criteria	_____
K) The organization has a policy preventing testing or a union contract prohibits testing	Use testing techniques that guarantee anonymity to check for understanding and satisfy that learning objectives are met	_____

training comparisons using tests are very common. An improvement in test scores shows the change in knowledge or skill of the participant that is attributed to the training. The principles of test development are similar to those for the design and development of questionnaires and attitude surveys.

The types of tests used in training can be classified in three ways. The first is based on the medium used for administering the test. The most common media for tests are written or keyboard tests; performance tests using simulated tools or actual equipment; and computer-based tests using computers and video displays. Knowledge and skills tests are usually written because performance tests are more costly to develop and administer. Computer-based tests and those using interactive video are gaining popularity. In these tests, a computer monitor or video screen presents the questions or situations, and participants respond by typing on a keyboard or touching the screen. Interactive videos have a strong element of realism because the person being tested can react to images, often moving pictures and video vignettes that reproduce the real job situation.

The second way to classify tests is by purpose and content. In this context, tests can be divided into aptitude tests or achievement tests. Aptitude tests measure basic skills or acquired capacity to learn a job. An achievement test assesses a person's knowledge or competence in a particular subject.

A third way to classify tests is by test design. The most common are objective tests, norm-referenced tests, criterion-referenced tests, essay tests, oral examinations, and performance tests. Objective tests have answers that are specific and precise, based on the objectives of a program. Attitudes, feelings, creativity, problem-solving processes, and other intangible skills and abilities cannot be measured accurately with objective tests. A more useful form of objective test is the criterion-referenced test. Oral examinations and essay tests have limited use in evaluating training; they are probably more useful in academic settings. The last two types of tests listed above are more common in training programs: criterion-referenced tests and performance testing evaluation. Both are described in more detail below.

A checklist for developing objective tests is presented as Figure 5.6.

Many objective tests are constructed as multiple choice tests. Figure 5.7 provides a guide for developing multiple choice items.

CRITERION-REFERENCED TESTS

The criterion-referenced test (CRT) is an objective test with a predetermined cutoff score. The CRT is a measure against carefully written objectives for the learning components of the program. In a CRT,

Figure 5.6. Downloadable checklist.

<div align="center">**Checklist for Developing Objective Tests**</div>
INTRODUCTION: Objective tests may be either knowledge-based or performance-based or both.
✔ A knowledge test measures the understanding of information or the achievement of an enabling skill. A knowledge test is either oral or written. Product knowledge and understanding the key components of a work process are examples of knowledge-based items.
✔ A performance test (sometimes referred to as skill application) requires that a participant perform a task (or tasks) that is performed on the job. Handling objections and the proper use of questioning techniques in a sales situation are examples of performance-based items.
✔ In either of the above situations, a participant being tested should have the same resources available that are available under actual conditions in the work setting. For example, if it is acceptable to use reference materials in the work setting, these materials should be available for use during the written or skill practice tests.
Step One—Review Course Objectives
✔ Carefully review the objectives to ensure that test items will reflect the behavior described by the objectives. Test items should not be written verbatim from objectives.
Step Two—Select the Proper Type of Test Item
✔ Select the type of test item that fits the objective and is best to use for evaluating the specific target audience. If the objective is performance-based, the test item should also be performance-based. A knowledge-base objective requires a knowledge-based test item.
✔ A performance-based test should be developed to include the task to be performed, the conditions under which the task will be performed,

Figure 5.6. (*continued*)

and the standards of acceptable performance. The job setting should be duplicated as closely as possible by using the same resources, equipment, materials, setting, and circumstances that will be encountered on the job.

Step Three—Write the Test Item

✔ Write test items for all objectives within each segment or module of instruction.

✔ Review the objectives to determine whether performance-based or knowledge-based tests are appropriate.

✔ Knowledge-based tests usually use a multiple choice test approach. This type of test usually measures a participant's comprehension of information presented in the training session. It requires the selection of a correct response from a number of possible answers. Refer to Figure 5.7 for the job aid on developing multiple choice tests.

✔ Performance-based tests should evaluate the process or product that is required in the work setting. The process refers to the steps the participant performs to accomplish the task. The product refers to the end result of the process. For example, probing is a step in the process used to uncover hidden objections in a sales situation. The end product occurs when the actual objections surface. It could be debated that the end product is not the surfacing of the objection, but the closing of the sale. Refer to figure 5.8 for the checklist on developing performance-based tests.

Step Four—Review Each Test Item

✔ Validity—Check that the test items are written so that the participant will perform the desired behavior under the conditions specified in the objectives and to the standard that is stated.

✔ The test item must be clear and unambiguous, free of opinion or interpretation.

✔ The test item must be written to the level of understanding and knowledge of the target population.

Figure 5.6. (*continued*)

✔ Avoid giving clues or hints to the correct answers.

✔ Avoid trick questions.

✔ Do not give the answer to one question by revealing it in another.

✔ Check closely for content and grammatical errors.

✔ Ensure that the test item is testing what should be known according to the program objectives.

Step Five—Match and Document the Test Items

✔ Document the match of test items to objectives. Document the correct answer for each test item.

Step Six—Design the Test

✔ Determine the type of test you will use and follow the guidelines below to design the test.

✔ Each objective should be tested.

✔ The number and range of test items should reflect the training time allocated to the objective.

✔ The test should differentiate between acceptable and unacceptable performance.

✔ The test should be a manageable length and should be able to be completed in the time allocated to it.

✔ Use a consistent approach to present test items. Do not change the style of question too frequently (i.e., changing style from true/false to multiple choice to ranking.)

✔ Do not split a test item from one page to another.

✔ Group and sequence test items by module.

✔ Provide clear directions to those who must take the test.

✔ Provide point values if they differ among the test items.

✔ Ensure that the test is a comprehensive reflection of the objectives and the required understanding and behavior.

Figure 5.7. Downloadable checklist.

Checklist for Developing Multiple Choice Test Items

INTRODUCTION: A multiple choice test item measures a participant's comprehension of information. When measured during or at the end of a session it is intended to measure the information presented during the training. When measured prior to the session, it is intended to measure comprehension prior to exposure to the training, sometimes called pre-testing. Multiple choice test items can measure a wide range of abilities, from simple recall to complex concepts. The participant is required to select a correct response from a number of possible answers.

Multiple choice test items are composed of two parts. The first part is called the "stem" and it asks the question. The second part contains the optional choices. Only one choice is correct, the others are called distractors.

Rules for Writing Multiple Choice Test Items

✔ Don't make the test item a test of reading ability.

✔ Do not use distractors in one item that provide clues or hints about answers for another item.

✔ When abbreviations or acronyms are used, define them the first time they are used.

✔ Write all test items needed for the same objective at the same time.

Rules for Writing the Stem

✔ The stem should include only one question. Use simple, direct, unambiguous language to clearly define the question.

✔ State the stem as a question, or as an incomplete sentence that is completed by one of the options, or as a description of a problem to be solved.

Figure 5.7. (*continued*).

✔ Avoid absolutes such as "always," "every," "all."

✔ If a word or phrase must be used in all the item choices, rephrase the stems to include it there.

✔ Use "Which of the following..." when there is more than one correct answer but only one of them is offered in the choices.

✔ Use "What..." when there is only one correct answer to the question (i.e., What is the final step of the customer complaint recovery process?).

✔ Ensure that there is only one correct answer included in the options.

✔ Do not use any variation of "all of the above" or "none of the above."

✔ Do not use similar terms or grammatical construction between the stem and answers that do not exist in other distractors. This can give clues to the correct answer.

✔ The distinction between the correct answers and the distractors should be fine enough to make selecting the answer difficult, but not so fine that it creates the possibility of more than one correct answer among three or four alternatives.

✔ Distractors should be as long, precise, and detailed as the correct answer.

✔ Present the options in a logical order:

  ■ shortest to longest, or largest to shortest

  ■ alphabetical order

  ■ ascending or descending numeric order

  ■ chronological order or reverse chronological order

✔ Ensure that the test items deal with important situations or requirements from the work setting. It should distinguish good job performers from poor performers.

Figure 5.7. (*continued*).

> ✔ Use the language of the work setting.
>
> ✔ It is usually best to offer four options.
>
> ✔ Use negative stems only when it is the best way or only way to ask a question, e.g., "not," "least," "except."

the interest lies in whether or not participants meet the desired minimum standards, not how that participant ranks with others. The primary concern is to measure, report, and analyze participant performance as it relates to the learning objectives.

Table 5.2 presents a reporting format based on criterion-referenced testing. This format helps explain how a CRT is applied to an evaluation effort. Four participants have completed a learning component with three measurable objectives that correspond to each of the modules. Actual test scores are reported, and the minimum standard is shown. For example, on the first objective, Participant 4 received a pass rating for a test which has no numerical value and which is simply rated pass or fail. The same participant met objective 2 with a score of 14 (10 was listed as the minimum passing score). The participant scored 88 on objective 3 but failed it because the standard was 90. Overall, Participant 4 satisfactorily completed the learning component. The column on the far right shows that the minimum passing standard for the program is at least two of the three objectives. Participant 4 achieved two objectives, the required minimum.

Criterion-referenced testing is a popular measurement tool sometimes used in training evaluation. The approach is helpful when it is necessary for a group of employees to learn new systems, procedures, or technology as part of a training program. Its use is becoming widespread. The process is frequently computer-based, making testing more convenient. It has the advantage of being objective-based, pre-

Table 5.2. Reporting format for CRT test data.

	OBJECTIVE 1 P/F	RAW SCORE	OBJECTIVE 2 STD	P/F	RAW SCORE	OBJECTIVE 3 STD	P/F	TOTAL OBJECTIVES PASSED	MINIMUM STANDARD	OVERALL SCORE
Participant 1	P	4	10	F	87	90	F	1	2 of 3	Fail
Participant 2	F	12	10	P	110	90	P	2	2 of 3	Pass
Participant 3	P	10	19	P	100	90	P	3	2 of 3	Pass
Participant 4	P	14	10	P	88	90	F	2	2 of 3	Pass
Totals 4	3 pass			3 pass			2 pass	8 pass		3 pass
	1 fail			1 fail			2 fail	4 fail		1 fail

cise, and relatively easy to administer. It requires clearly defined objectives that can be measured by tests.

PERFORMANCE TESTING

Performance testing allows the participant to exhibit a skill (and occasionally knowledge or attitudes) learned in a training initiative. The skill can be manual, verbal, or analytical, or a combination of the three. Performance testing is used frequently in job-related training where the participants are allowed to demonstrate what they have learned. In supervisory and management training, performance testing comes in the form of skill practices or role playing. Participants are asked to demonstrate discussion or problem-solving skills they have acquired. To illustrate the possibilities of performance testing, two examples are presented.

Example 1

Managers are participating in a training program that requires them to analyze the HR processes used in the organization. As part of the program, participants are given the project assignment to design and test a basic performance appraisal system. The instructional team observes participants as they check out their completed process, then carefully builds the same process and compares the results with those of the participants. These comparisons and the simulated performance of the design provide an evaluation of the project and represent an adequate reflection of the skills learned in the program.

Example 2

As part of a reorganization project, team members learn new products and sales strategies. Part of the evaluation requires team members to practice skills in an actual situation involving a sales presentation. Then participants are asked to conduct the skill practice on another member of the group using a real situation and applying the principles and steps learned in the project. The skill practice is observed by the trainer, and a written critique is provided at the end of the practice. These critiques provide part of the evaluation of the initiative.

For a performance test to be effective, the following steps are recommended in the design and administration of the test:

- The test should be a representative sample of the work/task related to the training initiative. The test should allow the participant to demonstrate as many skills taught in the program as possible. This increase the validity of the test and makes it more meaningful to the participant.

- The test should be thoroughly planned. Every phase of the test should be planned: the timing, the preparation of the participant, the collection of necessary materials and tools, and the evaluation of results.

- Thorough and consistent instructions are necessary. As with other tests, the quality of the instructions can influence the outcome of a performance test. All participants should be given the same instructions. They should be clear and concise. Charts, diagrams, blueprints, and other supporting information should be provided if they are normally provided in the work setting. If appropriate and feasible, the test should be demonstrated by the trainer so that participants observe how the skill is practiced.

- Procedures should be developed for objective evaluation, and acceptable standards must be developed for a performance test. Standards are sometimes difficult to develop because varying degrees of speed, skill, and quality are associated with individual outcomes. Predetermined standards must be developed so that employees know in advance what has to be accomplished to be considered satisfactory and acceptable for test completion.

- Information that may bias participants' reasoning should not be included. The learning module is included to develop a particular skill. Participants should not be led in this direction unless they face the same obstacles in the job environment.

With these general guidelines, performance tests can be utilized as effective tools for evaluations where demonstration of performance is a requirement of training evaluation. Although more costly than written tests, performance tests are essential in situations where a high degree of fidelity is required between work and test conditions. Figure 5.8 is a checklist for developing performance test items.

SIMULATIONS

Another technique for measuring learning is job simulation. This method involves the construction and application of a procedure or task that simulates the work involved in the training. The simulation is designed to represent, as closely as possible, the actual job situation. Participants try out their performance in the simulated activity and are evaluated based on how well the task is accomplished. Simulations may be used during the training program, at the end of the program, or as part of a follow-up evaluation.

Advantages of Simulations
Simulations offer several advantages for the trainer.

REPRODUCIBILITY
Simulations permit a job or part of a job to be reproduced in a manner almost identical to the real setting. Through careful planning and design, the simulation can have all of the central characteristics of the real situation. Even complex jobs, such as management, can be simulated adequately.

COST EFFECTIVENESS
Although sometimes expensive in the initial development, simulations can be cost effective in the long run. For example, it is cost prohibitive to train airline pilots to fly an airplane utilizing a $50 million aircraft. Therefore, an aircraft simulator is used to simulate all the flying conditions and enable the pilot to learn to fly before piloting an actual aircraft. If the cost of on-the-job learning is prohibitive, simulation is much more attractive.

Figure 5.8. Downloadable checklist.

Checklist for Developing Performance-Based Test Items

INTRODUCTION: A performance test item should evaluate the process or steps that the participant performs to accomplish a task or the end result. A performance test that evaluates a process is valuable for tasks where, if the process is not fully evaluated, much could be lost in the evaluation of the final product. For example, if a participant makes a mistake in the process, but the end result is correct, evaluators using a product-oriented test would not be aware of the mistake.

A performance test that evaluates a product looks at specific criteria that measure how well the participant performed. This is useful to evaluate tasks that can be performed in a number of different ways to achieve the desired outcome. In some instances, both the product and the process are evaluated.

Performance tests can be created as role plays in which two or more people portray a situation and one of the participants is evaluated by an observer.

The performance test should contain the following items:

✔ The objectives of the performance test must be stated. This can be a listing of the objectives or a brief description of what the participant must accomplish to successfully complete the performance test.

✔ Instructions to the evaluator should cover everything the evaluator needs to know or do to conduct the performance test.

  Instructions must be clear and detailed and must list all equipment, supplies, resources, etc. required to conduct the test.

  Additionally, the instructor should be provided with specific instructions for the participant and a checklist identifying exactly what performance steps are to be evaluated.

✔ A description of the instructional setting in which the test will be performed. This setting should match the conditions stated in the objective.

✔ Instructions to the participant that include what the participant must do. It should include any time limits, safety considerations, start and stop directions, and how the performance will be evaluated. The instructions should be clear to ensure that every participant is evaluated on the ability to perform the behavior stated in the objectives.

SAFETY CONSIDERATIONS

Another advantage of using simulations is safety, which is an important consideration when deciding on learning and testing methodologies. The safety component of many jobs requires participants to be trained in simulated conditions. For example, emergency medical technicians risk injury and even life if they do not learn emergency medical techniques prior to encountering a real-life situation. Firefighters are trained in simulated conditions prior to being exposed to actual fires. CIA agents are trained in simulated conditions before being exposed to their real-world environment.

Simulation Techniques

Several common simulation techniques are briefly described below.

ELECTRICAL/MECHANICAL SIMULATION

This technique uses a combination of electronics and mechanical devices to simulate real-life situations. It is used in conjunction with programs to develop operational and diagnostic skills. Examples of these types include simulated "patient," or a simulator for a nuclear power plant operator. Other less-expensive types of simulators have been developed to simulate equipment operation.

TASK SIMULATION

Another approach involves a participant's performance in a simulated task as part of an evaluation. For example, in an aircraft company, technicians are trained on the safe removal, handling, and installation of a radioactive source used in a nucleonic oil-quantity indicator gauge. These technicians attend a thorough training program on all of the procedures necessary for this important assignment. To become certified to perform this task, technicians are observed in a simulation, where they perform all the necessary steps on a checklist. After they have demonstrated that they possess the skills necessary for the safe performance of this assignment, they are certified by the instructor. This task simulation serves as the evaluation.

BUSINESS GAMES

Business games are simulations of a part or all of a business enterprise. Participants change the variables of the business and observe the effects of those changes. The participants are provided certain objectives, play the game, and have their output monitored. Their performance can usually be documented and measured. Typical objectives are to maximize profit, sales, market share, or return on investment. Those participants who maximize the objectives are those who usually have the highest performance in the program.

IN-BASKET

Another simulation technique called the "in-basket" is particularly useful in team leader, supervisory, and management training. The participant must decide how to deal with each of a series of items (e.g., memos, letters, reports) that represent items that normally appear in an in-basket. Time pressures often are added to create realistic conditions. The participant must decide what to do with each item (e.g., respond, pursue, delegate, refer, ignore) while taking into consideration the principles learned. The participant's performance is assessed. In some situations, every course of action for each item in the in-basket is rated, and a combination of the chosen alternatives provides an overall performance score.

CASE STUDIES

A perhaps less-effective but still-popular technique of simulation is a case study. A case study represents a detailed description of a problem and usually contains a list of several questions for the participant. The participant is asked to analyze the case and determine the best course of action. The problem should reflect the conditions in the real-world setting and the content in the training program.

The most common categories of case studies include:

- Exercise case studies, which provide an opportunity for participants to practice the application of specific procedures
- Situational case studies, which provide participants the opportunity to analyze information and make decisions regarding the particular situation

- Complex case studies, which are an extension of the situational case study, where the participant is required to process a large amount of data and information, some of which may be irrelevant
- Decision case studies, which require the participant to go a step further than the previous categories and present plans for solving a particular problem
- Critical-incident case studies, which provide the participant with a certain amount of information and withhold other information until it is requested by the participant.

Table 5.3. (Downloadable form.) Level of performance.

SKILL TO BE DEMONSTRATED	UNSATISFACTORY	NEEDS IMPROVEMENT	SATISFACTORY	ROLE MODEL
Use of questioning techniques	❑	❑	❑	❑
Uncovering hidden objections	❑	❑	❑	❑
Handling objections	❑	❑	❑	❑
Command of product knowledge	❑	❑	❑	❑
Communicating the core promotional message	❑	❑	❑	❑

- Action-maze case studies, which present a large case in a series of smaller units. The participant is required to predict at each stage what will happen next.

The difficulty in a case study lies in the objectivity of the evaluation of the participant's performance. Frequently, there can be many possible courses of action, some equally as effective as others, which makes it extremely difficult to obtain an objective, measurable performance rating for the analysis and interpretation of the case.

ROLE PLAYING/SKILL PRACTICE

In role playing, sometimes referred to as skill practice, participants practice a newly learned skill and are observed by other individuals. Participants are given their assigned roles with specific instructions, which sometimes include an ultimate course of action. The participant then practices the skill with other individuals to accomplish the desired objectives. This is intended to simulate the real-world setting to the greatest extent possible. Difficulty sometimes arises when other participants involved in the skill practice make the practice unrealistic by not reacting the way individuals would in an actual situation. To help overcome this obstacle, trained role players (nonparticipants trained for the role) may be used in all roles except that of the participant. This can possibly provide a more objective evaluation. The success of this technique also lies in the judgment of those observing the role playing. The skill of effective observation is as critical as the skill of the role player. Also, the success of this method depends on the participants' willingness to participate in and adjust to the planned role. If participant resistance is extremely high, the performance in the skill practice may not reflect the actual performance on the job. Nevertheless, these skill practices can be very useful, particularly in supervisory and sales training, to enable participants to practice discussion skills. Skill practice performance can be evaluated by an observer in one of two ways.

- **Go/No-Go Performance.** The observer can determine whether or not the participant has demonstrated the desired skill at the desired level of competence. There is no gray area. The participant either passes or fails.

- **Performance Scale.** A graduated scale can be used by the observer to determine the level of performance demonstrated by the participant. A scale similar to the one below can be used.

Figure 5.9 provides a checklist for designing role play scenarios.

Assessment Center Method

The final method for measuring learning with simulation is a formal procedure called the assessment center method. The feedback is provided by a group of specially trained observers called assessors. The assessment-center approach has been used for many years as an effective tool for employee selection. It now shows great promise as a tool for evaluating the effectiveness of a major learning module in a training program.

Assessment centers are not actually centers, such as locations or buildings. The term refers to a procedure for evaluating performance. In a typical assessment center, the individuals being assessed participate in a variety of exercises that enable them to demonstrate particular knowledge or skills, usually called "job dimensions." These dimensions are important to on-the-job success for the individuals involved in the training. It may take anywhere from four hours to three days for the participants to complete all the exercises. The assessors then combine ratings of each exercise for each dimension, removing subjectivity to reach a final rating for each participant.

In the assessment center process, a rating or "assessment" of the participants is given prior to the training (a pretest). After the training is implemented, the participants are assessed again to see if there are improvements in their performance within the job dimensions (a post-test). The use of a control group in an evaluation design helps produce evidence of the impact of the training.

Although the popularity of this method seems to be growing, it still may not be feasible for evaluating many training programs

Figure 5.9. Downloadable checklist.

Checklist for Designing A Role Play Scenario

✔ Describe the objectives of the role play. Describe what the role play participant should accomplish.

✔ Describe the environment surrounding the role play situation. Where the engagement is taking place, when it is occurring and the frame of reference of the parties involved.

✔ Provide information about the role being played (i.e., a customer). Who is this person, how has their day evolved thus far, what is their mind set? What is their interest in:

- the product

- the upcoming meeting

- the sales rep (or the customer)

- the knowledge possessed by the sale rep (or the customer)

- the needs of the sale rep

✔ Provide guidelines about how the role play participant should react to the approach, style and needs of the other party. Go for realism.

✔ Provide information about how the role play participant should behave and react regarding the key focus of the role play (i.e., closing a sale or overcoming an objection). Give ranges of behavior.

✔ Provide key questions that the role play participant should ask.

✔ Provide key answers that the role play participant should use to respond to key scenarios or situations.

---

because the use of an assessment center is quite involved and time consuming for the participants and the assessors. The assessors have to be carefully trained to be objective and reliable. However, for programs that represent large expenditures aimed at making improvements in the soft-data area, the assessment center approach may be the most promising way to measure the impact of the program. This is particularly true for an organization in which the assessment center process is already used for selection purposes.

In summary, there are many types of simulations that offer the opportunity for participants to practice what they have learned in a training program and to have their performance observed in simulated job conditions. They can provide extremely accurate evaluations if the performance in the simulation is objective and can be clearly measured.

Measuring learning with less-structured activities

In many situations, it is sufficient to have an informal check of learning to provide some assurance that participants have acquired the desired attitudes, knowledge, and/or skills. This approach is appropriate when other levels of evaluation are pursued. For example, if a level-3 application/implementation evaluation is planned, it might not be so critical to conduct a comprehensive level-2 evaluation, and an informal assessment of learning is usually sufficient. A comprehensive evaluation at each level is quite expensive. The following are some approaches to measuring learning when inexpensive, low-key, informal assessments are needed.

Exercises/Activities

Many training programs involve activities or exercises in which issues or problems must be explored, developed, or solved. Some of these are group activities, while others develop individual skills. When these tools are integrated into the learning activity, there are several ways in which to measure learning:

■ The results of an activity or exercise can be submitted for review and scoring by the instructor/facilitator. This becomes part of the participant's overall score for the program and becomes a measure of learning.

■ The results can be discussed in a group, with a comparison of the various approaches and solutions, and the group can reach an assessment of how much each individual has learned. This may not be practical in many settings, but can work in a few narrowly focused applications.

■ The solutions to the problem or exercises can be shared with the group, and each participant can provide a self-assessment indicating the degree to which the skills and/or knowledge have been obtained from the activity. This also serves as reinforcement in that participants quickly see the correct solution.

■ The trainer can review the individual progress of each participant to determine relative success. This is appropriate for small groups but can be very cumbersome and time consuming for larger groups.

Self-assessment

In many training programs, self-assessment may be appropriate. Participants are provided an opportunity to assess their acquisition of skills and knowledge. This is particularly applicable in cases where the higher-level evaluations are planned and it is important to know if actual learning is taking place. A few techniques can ensure that the process is effective:

■ The self-assessment should be made anonymously so that participants feel free to express realistic and accurate assessments of what they have learned.

■ The purpose of the self-assessment should be explained, along with the plans for use of the data. Specifically, if there are implications for program design or individual retesting, this should be discussed.

- If there has been no improvement or the self-assessment is unsatisfactory, there should be some explanation as to what that means and what the implications will be. This will help ensure that accurate and credible information is provided.

Trainer Assessment

A final technique is for the trainers to assess the learning that has taken place. Although this approach is very subjective, it may be appropriate when a higher-level evaluation is planned. One of the most effective ways to accomplish this is to provide a checklist of the specific skills that need to be acquired in the training. Trainers can then assess acquisition of particular knowledge or skills. If a particular body of knowledge needs to be acquired, a checklist of the categories should be developed for assuring that the individual has a good understanding of those items. Obviously, this could create a problem if the participants have not had the appropriate time and opportunity to demonstrate skills or knowledge acquisition, and the instructor may have a difficult time in providing appropriate responses. There is also the question of what to do if there is no evidence of learning. The specific consequences need to be considered and addressed before the method is used.

Administrative issues

There are several administrative issues that arise in measuring learning. Each is briefly discussed below and should be part of the overall plan for level-2 measurement.

CONSISTENCY

It is extremely important that different methodologies used for measuring learning are administered consistently from one group to another. This includes issues such as the time required to respond, the actual learning conditions in which the participants complete the process, the resources available to them, and the assistance from other members of the group. These issues can easily be addressed in the instructions.

MONITORING

In some situations, it is important for participants to be monitored as they are completing tests or other measurement processes. This ensures that each individual is working independently and also that someone is there to provide assistance or answer questions as needed. This may not be an issue in all situations but needs to be addressed in the overall plan.

SCORING

The scoring instructions need to be developed for the measurement process so that the person evaluating the responses will be objective in the process and provide consistent scores. Ideally, the potential bias from the individual scoring the instrument should be completely removed through proper scoring instructions and other information necessary to provide an objective evaluation.

REPORTING RESULTS

In some situations, participants are provided with assessment results immediately, particularly with self-scoring tests or with group-based scoring mechanisms. In other situations, the actual results may not be known until later. In these situations, a mechanism for providing scoring data should be built into the evaluation plan unless it has been predetermined that participants will not know the scores. What should not be done is to promise test scores and deliver them late or not at all.

The most common uses of level-2 data

There are four primary uses of Level-2 data, although other uses are possible. These four are:

Providing Individual Feedback to Build Confidence. Level-2 data, when provided directly to participants, provides reinforcement for correct responses and improves learning when responses are incorrect. This reinforces the learning process and provides much-needed feedback to participants in some situations.

Ensuring That Individual Learning Has Been Acquired. When it is desirable or necessary to verify that individual learning has occurred, instruments can be developed, validated, and implemented to test learning progress or outcomes. Many organizations shy away from individual testing of adults in a business setting due to the constraints mentioned earlier or due to restrictive labor contracts or other legal constraints.

Improving Training Programs. Perhaps the most important use of level-2 data is to improve the design of the training program. Consistently low responses in certain parts of a level-2 measure may indicate that inadequate coverage has been provided for a topic. Consistently low scores with all participants may indicate that the objectives and scope of coverage is too ambitious for the time allotted.

Evaluating Trainers. As with level-1 data, level-2 results are used to evaluate trainers and provide additional measures of the success of the training program. The trainer has a significant responsibility to ensure that the participants have learned the material (through proper design, delivery, and facilitation), and the level-2 measurement is a reflection of the degree to which the desired material has been acquired and internalized.

The Next Steps

The responsibility for collecting levels 1 and 2 evaluation data may rest with the trainer, a program coordinator, or the person who will be doing the evaluation. Early communication should take place between the evaluator and other parties to ensure that data requirements are met. When a follow-up evaluation is being conducted (level 3, 4, or 5), the evaluator should inform those with level-1 and 2 responsibilities that at least a summary of the level-1 and 2 data is needed to satisfy requirements to demonstrate a chain of impact.

FURTHER READING

Kaufman, Roger, Sivasailam Thiagarajan, and Paula MacGillis, editors. *The Guidebook for Performance Improvement: Working with Individuals and Organizations.* San Francisco: Jossey-Bass/Pfeiffer, 1997.

Kirkpatrick, Donald L. *Evaluating Training Programs: The Four Levels,* 2nd Edition. San Francisco: Berrett-Koehler Publishers, 1998.

Phillips, Jack J. *Handbook of Training Evaluation and Measurement Methods,* 3rd Edition. Houston: Gulf Publishing, 1997.

Phillips, Jack J. 1994, 1997. *Measuring The Return on Investment.* Vol 1 and Vol 2. Alexandria, VA: American Society for Training and Development.

Swanson, Richard A., and Elwood F. Holton III. *Results: How to Assess Performance, Learning, and Perceptions in Organizations.* San Francisco: Berrett-Koehler Publishers, 1999.

REFERENCES

1. Litwin, M. S. *How to Measure Survey Reliability and Validity.* Thousand Oaks, CA: Sage Publications, Inc., 1995.

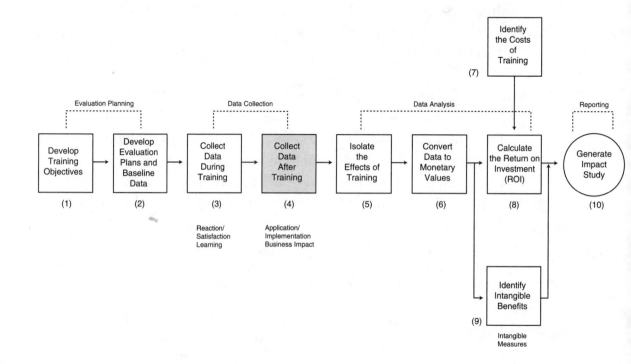

6

Step 4. Collect Data After Training (Levels 3 and 4)

The first follow-up step in the ROI process is collecting data after the training program has been conducted. This is the most important step; if data are not collected, the impact study cannot be completed. This step is also the most time-consuming and disruptive to the organization. The choice of method(s) to collect data after a program has been conducted depends on a variety of factors, such as the type of program, the willingness of the affected population to cooperate, the constraints posed by the organization, the availability of data, the cost of collecting the data, and the accuracy of the data. This chapter outlines ten common approaches for collecting post-program data, after the sources of data are defined. One or more of these approaches can be used to obtain credible data to complete the impact study.

THE BEST METHODS OF COLLECTING FOLLOW-UP DATA

The most common methods used to collect follow-up evaluation data are shown in Table 6.1. Some of these methods are suited to level-3 data; some are more suited to level-4 data; and some are suited to both levels. Data collection involves using feedback instruments, being mindful of timing and data sources. These methods are covered in more detail following Table 6.1.

Table 6.1. Post-program data-collection methods.

	LEVEL 3	LEVEL 4
1. Follow-up surveys	✔	
2. Follow-up questionnaires	✔	✔
3. Observations on the job	✔	
4. Follow-up interviews	✔	
5. Follow-up focus groups	✔	
6. Assignments related to the program	✔	✔
7. Action planning/improvement plans	✔	✔
8, Performance contracting	✔	✔
9, Program follow-up session	✔	✔
10. Performance monitoring		✔

Follow-up Surveys and Questionnaires. Attitude surveys differ from questionnaires. An *attitude survey* is an instrument that solicits opinions, beliefs, or values. A *questionnaire* may request an opinion, such as a reaction to the training program, but it also may cover a variety of other issues and use different types of questions. A questionnaire has much more flexibility and can elicit data ranging from attitudes to specific improvement statistics. Questions may seek level-4 data, such as asking about changes in sales or improvement in quality since the program was conducted, or may be in a multiple-choice or fill-in-the-blanks format. The questionnaire is the most common follow-up data-collection method. Ranging from brief assessment forms to detailed feedback tools, questionnaires can be used to obtain subjective information about skill application as well as to document measurable business results for an ROI analysis. The questionnaire is the preferred method for capturing Levels-3-and-4 data in many organizations.

A survey is a specific type of questionnaire, with several applications for measuring the success of training programs. The principles of survey construction and design are similar to questionnaire design. A checklist for questionnaire design is presented in Figure 6.1. This is a helpful job aid that can be referred to each time a new questionnaire is designed or one is revised. Some items may not apply in certain situations.

Refer to Figure 6.2, Questionnaire Example, and compare questions 1 through 21 to the checklist items 1 through 21 above. This will give you an idea of how to structure your questionnaires. Questions 8 through 12 are designed as level 4 and 5 business impact questions.

Participants should feel free to respond openly to questions without fear of reprisal. The confidentiality of their responses is of utmost importance, since there is usually a link between anonymity and accuracy. Therefore, questionnaires should be anonymous unless there are specific reasons why individuals have to be identified.

Observations on the Job. Observation is sometimes considered to be distasteful because it has the appearance of snooping on individuals. Despite this, it is still the best way to know exactly what behavior has taken place. It is most effective when the observer is undetected. An example of this is the professional shopper/customer used in customer-service training, who deals with the salesperson or customer-service representative that has been trained in order to monitor that person's sales presentation or customer-service skills.

Observation is enhanced with the use of a behavioral checklist, on which the predetermined behaviors sought in the observation are listed. The observation involves checking which behaviors are used and how often. Codes are used to link particular skills to behaviors, and the observer writes the codes on the form to describe the behaviors observed. With the delayed-report method, the observer does not have a notebook or checklist as he or she is observing, but breaks away frequently from the person being observed to record what has been observed. The advantage of this method is that the person being observed is more relaxed. The disadvantage is that some of the actual behavior could be lost because of the delay in recording.

Figure 6.1. Content issues checklist.

Follow-up Questionnaire: Content Issues Checklist

Item	Issue
1. Progress with Objectives	To what extent were the objectives met?
2. Action Plan Implementation	If an action plan was used, was it completed?
3. Relevance of Program	How relevant was the program?
4. Use of Materials	How useful were the materials while on the job?
5. Knowledge/Skill Enhancement	Which skills improved as a result of the training?
6. Skills Used	Which skills were most used?
7. Changes with Work	What changes were influenced by use of skills?
8. Improvements/Accomplishments	What business improvements resulted?
9. Monetary Impact	What is the monetary value of the improvements?
10. Confidence Level	What is the error possibility of any estimates?
11. Improvement Linked with Program	Percentage of improvement caused by program.
12. Investment Perception	Was the program worth the investment?
13. Linkage with Output Measures	Which important measures were influenced?
14. Other Benefits	What additional benefits resulted from program?
15. Barriers	What has inhibited performance in the work setting?
16. Enablers	What contributed to performance success?
17. Management Support	What additional management support is needed?
18. Other Solutions	What else might achieve the same results?
19. Recommendations for Target Audiences	Who else would benefit from this program?
20. Suggestions for Improvement	What improvements are needed in the program?
21. Other Comments	Additional comments or recommendations.

←— OPTIONAL (items 8–12)

Figure 6.2. Downloadable questionnaire example.

Impact Questionnaire for Leadership Development Program

*NOTE: This example is used only to illustrate
sample questions that may be asked on a follow-up questionnaire.
It is not intended to represent a document that is ready for implementation.*

Instructions

1. Please complete this questionnaire as promptly as possible and return it to the address shown on the last page. To provide responses, you will need to reflect on the Leadership Development Program and think about specific ways in which you have applied what you learned from each session. It may be helpful to review the materials from each session.

2. Please take your time as you provide responses. Accurate and complete responses are very important. You should be able to provide thorough responses in about 20 minutes.

3. You will need your action plan as you respond to several items on the questionnaire. Please review the action plan and make sure that each page is accurate and complete. Attach a copy of the action plan to the questionnaire when it is returned.

4. Please be objective in providing responses. In no way will your name be linked to your input. Your questionnaire and action plan will be viewed only by a representative from an outside firm, XX Company. Specific responses or comments related to any individual will not be communicated to your employer.

5. Your responses will help determine the impact of this program. In exchange for your participation in this evaluation, a copy of a report summarizing the success of the program will be distributed to you within a couple of months.

6. Should you need clarification or more information, please contact your trainer or a representative from XX Company.

Figure 6.2. Questionnaire example (*continued*).

Leadership Development Program Impact Questionnaire

Are you currently in a supervisory or management role?   Yes ❑   No ❑

1. Listed below are the objectives of the Leadership Program. After reflecting on the program, please indicate your degree of success in achieving these objectives. Please check the appropriate response beside each item.

Skill/Behavior	No Success	Very Little Success	Limited Success	Generally Successful	Completely Successful
A. Apply the 11-step goal-setting process	❑	❑	❑	❑	❑
B. Apply the 12-step leadership planning process	❑	❑	❑	❑	❑
C. Identify the 12 core competencies of outstanding leaders	❑	❑	❑	❑	❑
D. Identify 10 ways to create higher levels of employee loyalty and satisfaction	❑	❑	❑	❑	❑
E. Apply the concept of Deferred Judgment in five scenarios	❑	❑	❑	❑	❑
F. Apply the creative problem-solving process to an identified problem	❑	❑	❑	❑	❑
G. Identify the 7 best ways to build positive relationships	❑	❑	❑	❑	❑
H. Given a work-setting situation, apply the four-step approach to deal with errors	❑	❑	❑	❑	❑
I. Practice 6 ways to improve communication effectiveness	❑	❑	❑	❑	❑

2. Did you implement on-the-job action plans as part of the Leadership Program?   Yes ❑   No ❑

   If yes, complete and return your Action Plans with this questionnaire. If not, please explain why you did not complete your Action Plans._____

   _____

3. Please rate, on a scale of 1-5, the relevance of each of the program elements to your job, with (1) indicating no relevance, and (5) indicating very relevant.

	No Relevance		Some Relevance		Very Relevant
Group (Class) Discussions	1	2	3	4	5
Small Team	1	2	3	4	5
Skill Exercises (scenarios, role plays, etc.)	1	2	3	4	5
Program Content	1	2	3	4	5
Coaching and critique	1	2	3	4	5
Special Projects (leadership plan, job description, time log, money saving, etc.)	1	2	3	4	5

Figure 6.2. Questionnaire example (*continued*).

4. Have you used the written materials since you participated in
   the program?   Yes ❏   No ❏
   Please explain.
   _____

5. In the following result areas, please indicate your level of improvement during the last
   few months as influenced by your participation in the Leadership Program. *Check the*
   appropriate response beside each item.

Result Area	No Opportunity to Apply	No Change	Some Change	Moderate Change	Significant Change	Very Significant Change
A. ORGANIZING						
1) Prioritizing daily activities	❏	❏	❏	❏	❏	❏
2) Applying creative techniques	❏	❏	❏	❏	❏	❏
3) Organizing daily activities	❏	❏	❏	❏	❏	❏
4) Raising level of performance standards in area of responsibility	❏	❏	❏	❏	❏	❏
B. WORK CLIMATE						
1) Applying coaching	❏	❏	❏	❏	❏	❏
2) Applying techniques/initiatives that influence motivational climate	❏	❏	❏	❏	❏	❏
3) Implementing actions that influenced retaining people	❏	❏	❏	❏	❏	❏
4) Implementing job enrichment opportunities for valued associates	❏	❏	❏	❏	❏	❏
5) Implementing better control and monitoring systems	❏	❏	❏	❏	❏	❏
6) Applying techniques that influenced better teamwork	❏	❏	❏	❏	❏	❏
C. PERSONAL OUTCOMES						
1) Realizing improved written communications	❏	❏	❏	❏	❏	❏
2) Realizing improved oral communications	❏	❏	❏	❏	❏	❏
3) Realizing greater self-confidence	❏	❏	❏	❏	❏	❏
4) Working personal leadership plan	❏	❏	❏	❏	❏	❏

Figure 6.2. Questionnaire example (*continued*).

6. List the three (3) behaviors or skills from the above list that you have used most frequently as a result of the program.
   A) _____
   B) _____
   C) _____

7. What has changed about you or your work as a result of your participation in this program? (specific behavior change such as increased delegation to employees, improved communication with employees, employee participation in decision making, improved problem solving, etc.) _____
   _____
   _____

8. How has your organization benefited from your participation in the program? Please identify specific business accomplishments or improvements that you believe are linked to participation in this program; (Think about how the improvements actually resulted in influencing business measures such as increased revenue, increased overall shipments, improved customer satisfaction, improved employee satisfaction, decreased costs, saved time, etc.)
   _____
   _____
   _____
   _____

9. Reflect on your specific business accomplishments/improvements as stated above and think of specific ways that you can convert your accomplishments into a monetary value. Along with the monetary value, please indicate your basis for the calculations.
   Estimated monetary amount $_____
   Indicate if above amount is weekly, monthly, quarterly, or annually.
   ❑ Weekly   ❑ Monthly   ❑ Quarterly   ❑ Annually
   What is your basis for your estimates? (What influenced the benefits/savings and how did you arrive at the value above)?
   _____
   _____
   _____
   _____
   _____
   _____

Figure 6.2. Questionnaire example (*continued*).

10. What level of confidence do you place on the above estimations?

    _____% Confidence (0% = No Confidence, and 100% = Certainty)

11. What percentage of the improvement above was actually influenced by the application of knowledge and skills from the *Leadership Program*?

    _____% Confidence (0% = None, and 100% = All)

12. Do you think the *Leadership Program* represented a good investment for your organization?

    ❏ Yes   ❏ No

    Please explain.

    _____

    _____

13. Indicate the extent to which you think your application of knowledge, skills, and behavior learned from the Leadership Program had a positive influence on the following business measures in your own work or your work unit. Please check the appropriate response beside each measure.

Business Measure	Not Applicable	Applies But No Influence	Some Influence	Moderate Influence	Significant Influence	Very Significant Influence
A. Work output	❏	❏	❏	❏	❏	❏
C. Cost control	❏	❏	❏	❏	❏	❏
D. Efficiency	❏	❏	❏	❏	❏	❏
E. Response time to Customers	❏	❏	❏	❏	❏	❏
F. Cycle time of products	❏	❏	❏	❏	❏	❏
G. Sales	❏	❏	❏	❏	❏	❏
H. Employee turnover	❏	❏	❏	❏	❏	❏
I. Employee absenteeism	❏	❏	❏	❏	❏	❏
J. Employee satisfaction	❏	❏	❏	❏	❏	❏
K. Employee complaints	❏	❏	❏	❏	❏	❏
L. Customer satisfaction	❏	❏	❏	❏	❏	❏
M. Customer complaints	❏	❏	❏	❏	❏	❏
N. Other (please specify)	❏	❏	❏	❏	❏	❏

Please cite specific examples or provide more details:

_____

_____

_____

_____

Figure 6.2. Questionnaire example (*continued*).

14. What additional benefits have been derived from this program?

    _____

    _____

15. What barriers, if any, have you encountered that have prevented you from using skills/behaviors gained in the Leadership Program? Check all that apply.
    ❏ I have had no opportunity to use the skills.
    ❏ I have not had enough time to apply the skills.
    ❏ My work environment does not support the use of these skills/behaviors.
    ❏ My supervisor does not support this type of program.
    ❏ This material does not apply to my job situation.
    ❏ Other (please specify):

    _____

    If any of the above are checked, please explain if possible. _____

    _____

    _____

16. What enablers, if any, are present to help you use the skills or knowledge gained from this program? Please explain.

    _____

    _____

    _____

17. What additional support could be provided by management that would influence your ability to apply the skills and knowledge learned from the program?

    _____

    _____

    _____

18. What additional solutions do you recommend that would help to achieve the same business results that the *Leadership Program* has influenced? ?

    _____

    _____

    _____

Figure 6.2. Questionnaire example (*continued*).

19. Would you recommend the Leadership Program to others?   Yes ❑   No ❑

    Please explain. If no, why not. If yes, what groups/jobs and why?

    _____

    _____

    _____

20. What specific suggestions do you have for improving this program?

    _____

    _____

    _____

21. Other Comments:

    _____

    _____

    _____

    Date of Last Session: _____

    +---+
    | Please return completed questionnaire and action plan |
    | in the enclosed envelope or mail directly to: |
    | |
    | |
    +---+

Another method of observation is video recording. Video cameras are placed on individuals to record what behavior is occurring after the training. Audio monitoring is used to tape record individual conversations to determine if specific skills are used. This is most often used in customer-service training where telephone calls are monitored to see the extent to which employees are using desired skills. A final option is computer monitoring of activity to keep track of, for example, how long individuals are away from the desk.

Follow-up Interviews. Generally, the interview is more thorough than the questionnaire and can give more complete data because you can probe and make sure the information is complete and the participant fully understands the question. It also has the unique advantage of identifying a potential success story. When significant data is uncovered in the interview, it can be probed to determine details that could be reported as a significant success from the program. The disadvantage of the interview is that it is expensive and time-consuming for the training staff, interviewer, and participant. Also, the data tabulation, summary, and analysis may be difficult because some data are subjective.

Follow-up Focus Groups. The focus group has an advantage over the interview in that it is a more economical approach. It takes less time and has the synergy of the group discussion (i.e., a point made by one individual may be amplified by another.) For example, one individual may indicate a specific change implemented after completing a training program. Another individual perhaps had not thought about that issue but realizes that he or she has made the same change and got similar results. Without the focus group, this additional piece of evidence would not necessarily have been generated. The focus group works best with soft-skill data, but it has the disadvantage of not being able to probe for details because of the number of individuals in the group. Also, sometimes the group will inhibit some discussion that might otherwise flow from an interview. Focus groups should be conducted by skilled facilitators in order to effectively get data from a group.

Assignments Related to the Training. In some cases, follow-up assignments can develop level-3 and level-4 data. In a typical follow-up assignment, the participant is instructed to meet a goal or complete a particular task or project by the determined follow-up date. Participants take their knowledge and skills, apply them to the job, and report what happened in a predetermined format. (For example, in skill building with supervisors, participants may be asked to apply the skills following the steps taught and report back their particular application, its success, and its outcome. A summary of the results of these completed assignments provides further evidence of the impact of the program

Action Planning/Improvement Plans. It is critical to understand what goes into the action plan if it is to be used as a data collection method. Figure 6.3 includes one of many forms that can be used with an action planning process. Figures 6.4 and 6.5 show the same form with an example that is used with a leadership program. One advantage of action planning is that a participant can implement several plans. (For example, if one variable focuses on quality improvement, another on absenteeism, and yet another on customer satisfaction, there can be a different plan for each). Another advantage is that participants who work in different organizational units or even different companies can use the process to develop data on variables unique to their situations and jobs.

The action plan is the most common type of follow-up assignment. Participants are required to develop action plans as part of the training program. Action plans contain detailed steps to accomplish specific objectives related to the program. The action plan shows what is to be done, by whom, and the date by which the objectives should be accomplished. The action plan approach is a straightforward, easy-to-use method for determining how participants will change their behavior on the job and achieve success with the training. Action plans, as used in this context, do not require the prior approval or input from the participant's supervisor, although it may be helpful. The approach produces data that answer such questions as:

Figure 6.3. Downloadable action plan worksheet.

Action Plan for the
_____*Training Program*

Name _____   **Job Title** _____
Department _____   **Organization** _____
Training Start Date _____   **Follow-up Date** _____
Trainer_____

Note: This Action Plan document should be initiated during the training process. Part I and Part II will be completed at different times. The trainer will instruct you on how to proceed, provide you with examples of properly completed action plans, and answer questions you may have about your plan. Please be thorough and accurate with the information. Three Action Plan forms are included, one for each major improvement area that you identify. If additional areas are identified, please request another Action Plan document. The completed Action Plan should be returned a few months after you compete the process.

Instructions

Part I, General Information. The entire Action Planning Process should be reviewed with you early in the training process. The idea is for you to set an objective(s) for improvement in your employer's work setting. During the training program, you will set the objective(s) and determine the planning steps. You will implement the plan over a four-to-five-month period in your work setting. At the end of the four-to-five-month period, you will be asked to return your completed action plan so that your documented achievements can be consolidated with your classmates' achievements to determine a return on investment for the training program.

The exact timing for completion of Part I is determined by your instructor depending upon circumstances with your class. In any event, Part I of the plan should be completed and reviewed by your instructor no later than the last class session. At the top of the form, include your name and your *Objective*, such as to increase sales or productivity. Other examples might be to improve the quality of your work or the work completed by your team in your work setting. Your instructor will provide you with examples of plans that have been completed and used for documenting your achievements and calculating return on investment.

The follow-up date to collect the plans is usually four to five months after the last session. The *Evaluation Period* ranges from the time that Session One takes place to the date when the Action Plan is returned with completed information. The *Improvement Measure* is the specific measure of success, such as sales, time-savings, or reduction in errors or rework, that will determine the achievement of your objective. The *Current Performance* is the level of performance prior to initiating the Action Plan. The *Target Performance* is your goal for improving this particular measure. The instructor should sign the instructor signature block after he/she has reviewed Part I of the Action Plan.

Figure 6.3. Action plan worksheet (*continued*).

Part I Specific Steps. On the left side of the form, specific action steps should be included. These are the specific steps, behaviors, or tasks that will be implemented, changed, or improved as you work toward your objective and improvement measure.

Part I End Result. On the right side of the form, list the end result or consequences that you expect from each of the specific steps.

Expected Intangible Benefits. List the intangible benefits that you expect from the training. These are benefits that are not necessarily quantifiable or convertible to monetary values. For example, these might be things like improved morale, increased communication, or reduction in conflicts.

Part II Analysis. The most important part of this form is the information included in the analysis. This part of your plan is completed during an approximate three-month period after you complete the training. The *Improvement Measure* is repeated from Part I.

- **Question A** refers to the particular unit of measure. Two examples are provided below. If "sales" is the improvement measure, then the unit of measure is "one item sold." If "time savings" is the improvement measure, then the unit of measure is "hours saved" or "days saved," etc.

- **Question B** is the value of one unit if it is improved, eliminated, increased, or enhanced. Two examples are provided below.
  Sales—If one additional item is sold, what is it worth in added value? This may require estimations or input from experts in the organization.
  Time Savings—If you saved time and the unit of measure is "hours saved," then how much is an hour worth? Time saved is usually valued at the hourly wage rate plus fringe benefits.

- **Question C** provides space to indicate how you arrived at the value. For example, what assumptions were made in developing the value? Who provided the data if someone other than yourself? Is it a standard value used in your organization or is it a value you determined?

- **Question D** refers to the amount of the change during the evaluation period, comparing the monthly value for the month prior to initiation of the Action Plan to the most recent month in the follow-up period. You should provide this as a monthly value. Two examples are provided below.
  Sales—If total sales were 15 units sold per month at the beginning of the Action Plan period and 20 units per month sold at the end of the evaluation period, then the increase is 5 units.
  Time Savings—If you (or members of your work team) had 20 hours of unproductive time per month prior to initiating the Action Plan, and the training influenced an improvement so that after the training you now have only 6 hours of unproductive time per month, then the time savings is 14 hours per month.

After you have provided the amount of change above, give an explanation of what you or your employee work team did to cause the results to happen and how the training program played a part in the achievement. Please be as specific as possible. Use the back of the form or additional attachments if necessary.

Figure 6.3. Action plan worksheet (*continued*).

- **Question E** provides an opportunity to indicate the subjective nature of this process. Indicate the confidence you have in the above information, with 100% indicating certainty and 0% meaning no confidence. Review all of the questions under the analysis section and indicate the level of confidence you have in the information you have provided, including estimates that you have made.

- **Question F** asks for the amount of the change that was actually caused as a result of you or your team applying knowledge or skills learned from the training program. Here you should consider the other factors that could have influenced changes in the measure. Think about the influence of your actions from the training program and then estimate, with a percentage, how much of the change was actually related to the training, from 0% to 100%.

- **Question G** asks you to estimate the percentage of "time saved" (if time-savings is your measure) that was then used in productive ways. Time saved that is not used productively is not considered an improvement. Think about the time you (or your team) saved and how you (or team members) used that time. Estimate the percentage used in productive ways from 0% to 100%. If time-savings is not your improvement measure, you should leave this question blank.

Actual Intangible Benefits. List actual intangible benefits that were realized from your achievements. Think about the things that you and your employee work team are doing differently to benefit your work situation and your organization.

Figure 6.3. Action plan worksheet (*continued*).

Worksheet Part I—Action Plan For The _____ *Training Program*

Name: _____ Instructor Signature: _____ Follow-up Date: _____

Objective: _____ Evaluation Period: _____

Improvement Measure: _____ Current Performance: _____ to _____

Target Performance: _____

SPECIFIC STEPS: I will do this ➡

1. _____

2. _____

3. _____

4. _____

5. _____

6. _____

7. _____

END RESULT: So that ➡

1. _____

2. _____

3. _____

4. _____

5. _____

6. _____

7. _____

EXPECTED INTANGIBLE BENEFITS ➡

_____

_____

_____

Figure 6.3. Action plan worksheet (*continued*).

Worksheet Part II—Action Plan For The _____ *Training Program*

Name: _____ Objective: _____

Improvement Measure: _____ Current Performance: _____ Target Performance: _____

ANALYSIS

A. What is the unit of measure? _____ Does this measure reflect your performance alone? Yes ☐ No ☐

If not, how many employees are represented in the measure? _____

B. What is the value (or cost) of one unit? $ _____

C. How did you arrive at this value? _____

_____

D. How much did this measure change during the last month of the evaluation period compared to the average before the training process?

(monthly value) _____ Please explain the basis of this change and what you or your team did to cause it.

_____

E. What level of confidence do you place on the above information? (100% = Certainty and 0% = No Confidence) _____ %

F. What percentage of this change was actually caused by the application of the skills from the _____ Training _____ %
Process (0% to 100%)

G. If your measure is time-savings, what percentage of this time saved was actually applied toward productive tasks? (0% to 100%) _____ %

ACTUAL INTANGIBLE BENEFITS

_____

_____

_____

- What measures are being targeted?
- What is the unit of measurement?
- What is the current and targeted performance?
- What on-the-job improvements have been realized since the program was conducted?
- How are the improvements linked to the program?

Following the instructions in Figure 6.3 and the examples in Figures 6.4 and 6.5, participants should review how the action planning process may apply to their situation.

Performance Contracting. Performance contracting is a variation of the action-planning process. When preprogram commitments are made between the participants, trainer, and supervisor of the participant, the action plan then becomes a performance contract. This means there is a contract to improve performance. Typically the supervisor and participant meet prior to the training to develop the performance plan and contract. They then meet several times during and after the training for the purpose of reporting, feedback, and coaching.

It is important for the variables selected to be connected with the programming in some way, and the skills, steps, and techniques used are taken out of the material from the training program. These are the specific steps involved in a performance contract. Because of the high level of management involvement and the prework commitment that is often required, this may be reserved for those programs that require more resources in terms of money or time (e.g., a two- or three-week program).

Follow-up Session. Programs redesigned to have multiple sessions instead of one continuous series of days are becoming more commonplace. (E.g., a two-week program is redesigned into five-day and three-day segments and the organization is able to cover the same material. A three-month period is inserted between the two segments so that participants can apply the skills taught during the first week.) This is important from an evaluation standpoint in that follow-up sessions require a great deal of time devoted to evaluation data, where

Figure 6.4. Downloadable example action plan: Sample A.

Worksheet Part I—Action Plan For The _____ Training Program

Name: _Medicine Gelatin Manager_   Instructor Signature: _Amelia Lora_   Follow-up Date: _____

Objective: _Elimination of Gel Waste_   Evaluation Period: _January_ to _May_

Improvement Measure: _Quality_   Current Performance: _8,000 kg's waste monthly_ Target Performance: _Reduce waste by 80%_

SPECIFIC STEPS: I will do this ▶

1. Take a more active role in daily gelatin schedule to ensure the manufacture and processing control of gelatin quantities.

2. Inform supervisors and technicians on the value of gelatin and make them aware of waste.

3. Be proactive to gelatin issues before they become a problem.

4. Constantly monitor hours of encapsulation lines on all shifts to reduce downtime and eliminate the possibility of leftover batches.

5. Provide constant feedback to all in the department including encaps machine operators.

END RESULT: So that ▶

1. Better control of gelatin production on a daily basis. This will eliminate the making of excess gelatin which could be waste.

2. Charts and graphs with dollar values of waste will be provided to give awareness and a better understanding of the true value of waste.

3. Able to make gelatin for encapsulation lines and making better decisions on the amounts.

4. Eliminate the excess manufacturing of gelatin mass and the probability of leftover medicine batches.

5. Elimination of unnecessary gelatin mass waste.

EXPECTED INTANGIBLE BENEFITS ▶

Gel mass will decrease to a minimum over time, which will contribute to great financial gains for our company (material variance), which will put dollars into the bottom line.

Figure 6.4. Example action plan: Sample A (*continued*).

Worksheet Part II—Action Plan For The _____ Training Program

Name: _Medicine Gelatin Manager_   Objective: _Amelia Lora_

Improvement Measure: _Quality_   Current Performance: _8,000 kg's waste monthly_ Target Performance: _Reduce waste by 80%_

ANALYSIS

A. What is the unit of measure? _Waste reduction_   Does this measure reflect your performance alone?   Yes ☐   No ☑

B. If not, how many employees are represented in the measure? ___ 32 ___

C. What is the value (or cost) of one unit? $ _3.60 per kilogram of gelatin mass_

How did you arrive at this value?

This is the cost of raw materials and is the value we use for waste.

D. How much did this measure change during the last month of the evaluation period compared to the average before the training process?

(monthly value) _Currently 2,000 kg's monthly waste_   Please explain the basis of this change and what you or your team did to cause it.

6,000 kilograms of waste eliminated. Reducing machines from 19 to 12 created additional savings, but did not calculate. Gains in machine hours
efficiency in Encaps Department. More awareness of gel mass waste and its costs. Key contributing factors were problem-solving skills, communicating
with my supervisors and technicians and their willing response, and my ability to manage the results.

E. What level of confidence do you place on the above information? (100% = Certainty and 0% = No Confidence) ___ 70 ___ %

F. What percentage of this change was actually caused by the application of the skills from the _____ Training ___ 20 ___ %

Process (0% to 100%)

G. If you measure is time-savings, what percentage of this time saved was actually applied toward productive tasks? (0% to 100%) ___ NA ___ %

ACTUAL INTANGIBLE BENEFITS

Figure 6.5. Downloadable example action plan: Sample B.

Worksheet Part I—Action Plan For The _____ Training Program

Name: _Bill Burgess, Senior Engineer_   Instructor Signature: _Amelia Lora_   Follow-up Date: _Oct 13_

Objective: _Enable my team to sustain continuous improvement by end of next QTR._   Evaluation Period: _Jul 13_ to _Oct 13_

Improvement Measure: _Time savings_   Current Performance: _32 hours unproductive time per month_   Target Performance: _Zero unproductive hours_

SPECIFIC STEPS: I will do this ▶

1. _Convince the technicians on my team that I am serious about continuous improvement._

2. _Get my team involved in identifying key result areas and related improvement needs and focus on improving these._

3. _Provide coaching to team members to help them break paradigms._

4. _Consistently apply the principle of deferred judgment._

5. _Apply the 5-step approach to handling mistakes._

6. _Teach my team members to understand and diagnose before prescribing._

7. _Continuously seek ways to inspire ownership and involvement in team members._

8. _Observe & think about ways team members respond to my behavior and approach. Document improvements like time savings, quality improvement, output improvement, cost savings, and other beneficial changes. Measure the results objectively, provide feedback & recognition to team members._

END RESULT: So that ▶

1. _Team members understand my expectations and intentions._

2. _Team members become committed and feel empowered to act._

3. _Team members see new approaches, possibilities, solutions._

4. _Ideas get a fair hearing & team members get an opportunity to support them and find ways to demonstrate their practicality._

5. _Support risk taking, learning accountability & growth occurs._

6. _Team members minimize errors and rework by addressing problems and not symptoms._

7. _Get team members excited about their achievements._

8. _I can learn what works and what doesn't, I can close the feedback loop so that the team internalizes that achievements are rewarded and I can begin to put a dollar value on the results of my efforts._

EXPECTED INTANGIBLE BENEFITS ▶

People will feel a greater sense of my desire to see them achieve. I will become more satisfied with my own achievement and my team will function more like a team. My team will make a greater contribution to the company's goals. Individual team members will become more fulfilled. This can also lead to improved employee satisfaction and improved customer service.

Figure 6.5. Example action plan: Sample B (continued).

Worksheet Part II—Action Plan For The _____ Training Program

Name: _Bill Burgess_   Objective: _To enable my team to sustain continuous improvement by end of next quarter._

Improvement Measure: _Time savings_   Current Performance:_32 hours unproductive time per month_   Target Performance: _Zero unproductive hours_

ANALYSIS

A. What is the unit of measure? _Team hours saved per month._   Does this measure reflect your performance alone?   Yes ☐   No ☑

     If not, how many employees are represented in the measure? _Eight_

B. What is the value (or cost) of one unit? $ _52.00 for each hour saved._

C. How did you arrive at this value?

     Team member salary is $40.00 per hour. I added 30 percent for benefits.

     _30% of $40 = $12_   _$40 + 12 = $52 per hour._

D. How much did this measure change during the last month of the evaluation period compared to the average before the training process?

     (monthly value) _Improved from 32 hours of unproductive time per month to only 8 hours for the entire team._

     Please explain the basis of this change and what you or your team did to cause it. _Improvement in unproductive time. Got my team involved in identifying_

     key result areas and improvement needs and they came through. My ability to listen and not be judgmental played a key part in the results they achieved.

     Each hour saved due to increased productivity (resulting in less unproductive time) is credited to the program. 100% of the hours saved has been used in

     productive ways to benefit the company, and most of what was applied to achieve these results was as a result of what I have learned from the training program.

E. What level of confidence do you place on the above information? (100% = Certainty and 0% = No Confidence)   _95_ %

F. What percentage of this change was actually caused by the application of the skills from the   Training   _75_ %

     Process (0% to 100%)

G. If your measure is time-savings, what percentage of this time saved was actually applied toward productive tasks? (0% to 100%) _100_ %

ACTUAL INTANGIBLE BENEFITS

Because of my new approach with my team, they respond better by taking the initiative and working together to achieve results. As I continue to practice

what I learned from the training, my confidence increases and I will be able to accomplish more and feel better about myself.

individuals report what they have accomplished with the program as well as the results that they obtained. An additional educational component is often added where advanced skills are covered or particular techniques are reviewed to try to remove identified barriers to success.

Performance Monitoring. Monitoring performance data enables management to track performance in areas such as output, quality, costs, customer satisfaction, employee satisfaction, and time allocated to activities. Data are available in every organization to measure performance. In determining the use of data in the evaluation, the first consideration should be to use existing databases and reports. In most organizations, performance data suitable for measuring the improvement resulting from training programs are available. If not, additional record-keeping systems may have to be developed for measurement and analysis. When considering the development of new record-keeping systems, economics enters the picture. Is it economical to develop the record-keeping system necessary to evaluate a training program? If the costs are greater than the expected return for the entire program, then it is meaningless to develop the system. Under these circumstances, other methods should be explored for feasibility and acceptance by stakeholders.

FINDING THE MOST RELIABLE DATA SOURCES

The availability of data is the first major issue faced by the evaluator. Several sources are available for level-4 (business impact) data that provide evidence of a training program's success.

Organizational Performance Records. The most credible data sources are the records and reports of the organization. Records that reflect performance in a work unit, department, division, region, or company can be individual or group based. Collecting data from this source is preferred for level-4 evaluation, since it usually reflects business impact data and it is relatively easy to obtain. However, record keeping is not always precise or consistent in some organizations, and this may make it difficult to locate or use the data.

Participants. Participants are the most widely used source and they are frequently in a position to provide rich data. Participants are asked how skills and knowledge, acquired in a program, have been applied on the job. When level-4 data is sought, they are asked to explain the impact of their applications. They are very credible, since they are the individuals who have achieved the performance and are often the most knowledgeable about the processes and other influencing factors. It is sometimes difficult to capture participant data in a consistent manner.

Supervisors of Participants. Those people who directly supervise or lead program participants are an excellent source for certain types of data. In many situations, they observe the participants as they attempt to use the knowledge and skills acquired in the program. Consequently, they can report on the successes linked to the program as well as the difficulties associated with job application. Although supervisor input is usually best for level-3 data, it can be useful for level-4 data.

Subordinates of Participants. When supervisors and managers are being trained, such as in leadership programs, their subordinates can provide information about the perceived changes in observable behavior that have occurred since the program was conducted. While subordinates can observe a change in performance, it is difficult for them to determine the reasons for the change. Input from subordinates is appropriate for level-3 data but not level-4 data, since they may not be knowledgeable about the business impact. Collecting data from subordinates is often avoided because of the potential biases that can enter into the feedback process.

Team/Peer Group. For some types of programs, individuals who serve as team members with the participant or occupy peer-level positions in the organization are a source of data. In these situations, peer group members provide input on perceived behavioral changes of participants. This source of data is more appropriate when all team members participate in the program and they report on the collective efforts of the group or behavioral change of specific individuals. Because of the subjective nature of this process, and the lack of opportunity to fully evaluate the application of skills, this source of data is somewhat limited.

Internal/External Groups. In some situations, internal or external groups, such as the training and development staff, subject-matter experts, and external consultants, may provide input on the success of the participants as they apply the skills and knowledge acquired in the program. Data from this source has limited use in some cases. Because internal groups may have a vested interest in the outcome of evaluation, their input may lack credibility. Input from external groups is appropriate with certain types of observations of on-the-job performance, but feedback from casual observations should be avoided.

QUESTIONS TO ASK IN ANY TYPE OF FOLLOW-UP EVALUATION

Performance data such as output (sales, production, shipments, loans processed), quality improvements, time saved, cost savings, customer satisfaction, etc., are best obtained from organizational records. However, this type of data is not always available from organizational records, or the data is only kept at organizational-unit levels and is not available on individuals. A useful and credible alternative is to collect the data from individual participants or others close to the situation. This can be accomplished through questionnaires, interviews, focus groups, action plans, or performance-contracting instruments. When collecting data from individuals or groups, the series of questions below can be asked to yield level-3 and -4 data.

1. How did you apply what you learned in this program?

2. What was the impact of these efforts in your work unit?
3. What measures were changed in your work unit?
4. How much did they change?
5. What is the monetary value of the changes?
6. How did you arrive at this value?
7. What percent of this improvement was actually caused by the training?
8. What level of confidence do you place on this value? (Expressed as a percentage).

These questions elicit the data that we often need. They also introduce the question of how much of the improvement was actually caused by the training program. This is one way of isolating the effects of the training, which will be covered in Chapter 7. This approach is further illustrated in Figure 6.2. Questions 7, 8, 9, 10, and 11 address these level-3 and -4 issues. Figures 6.3, 6.4, and 6.5 include the same questions to capture business impact data.

Example calculation using estimates for confidence level and attribution. Using the data from Figure 6.4, the following is the proper calculation using the principles of analysis described above.

Value of one kilogram of waste (item B): $3.60
Kilograms of waste eliminated (item D): 6,000 kilograms
Confidence level of estimate (item E): 70%
Improvement attributed to training program (item F): 20%
B x D ($3.60 × 6000) = $21,600
$21,600 × 0.70 confidence level = $15,120
$15,120 × 0.20 = $3,024 attributed to training program

IMPROVING RESPONSE RATES TO QUESTIONNAIRES

Practical steps can be taken to improve response rates to questionnaires. Response rates should generally exceed 50 percent to have good representation of the group being surveyed and to feel comfortable with the completeness of the data. With small numbers,

Figure 6.6. Checklist for improving the response rate for
questionnaires and action plans.

✔ If appropriate, such as with action planning, provide advance communication on the process.

✔ Clearly communicate the reason for the questionnaire or action plan.

✔ Indicate who will see the results of the questionnaire or action plan.

✔ Show how the data will be integrated with other data.

✔ Keep the instrument simple and brief.

✔ Discuss the instrument during the training program. (This is imperative with action plans.)

✔ Estimate the time needed to complete questionnaire.

✔ Make input anonymous.

✔ Make it easy to respond (e.g., include a self-addressed stamped envelope).

✔ Use the local manager to help distribute the instrument and show support.

✔ If appropriate, let the target audience know that they are part of a carefully selected sample.

✔ Use one or two follow-up reminders.

✔ Have the introduction letter signed by a top executive.

✔ Consider incentives for completing and returning the forms.

✔ Send a summary of results to target audience.

less than 50 people, it is desirable to have a 100-percent response. A review of Figure 6.6, Checklist for Improving Response Rates, will be useful to help increase the amount and quality of data you receive.

GETTING STARTED WITH DATA COLLECTION

Use the tool presented in Figure 6.7, Personalized Action Steps, to plan your own data collection efforts. Determine issues you need to tackle or problems that you must overcome in your data collection initiatives. Think about who is responsible to address each issue and who can provide assistance in addressing it. For example, if your organization is not open to the use of questionnaires, list that as an issue and then take the responsibility to find out why by listing your name in the column labeled, "responsible person." Then determine who can help resolve the issue. Perhaps several members of senior management are key players. Maybe the issue exists only because no one has challenged it in recent years. Think about who the thought leaders are in your organization. Select a few of these people to interview, and then develop a plan for the interviews that will enable you to succeed in getting a favorable decision.

FURTHER READING

Broad, M.L. and Newstrom, J.W., *Transfer of Training*. Reading, MA: Addison-Wesley, 1992.

Kaufman, Roger, Sivasailam Thiagarajan, and Paula MacGillis, editors. *The Guidebook for Performance Improvement: Working with Individuals and Organizations*. San Francisco: Jossey-Bass/Pfeiffer, 1997.

Kirkpatrick, Donald L. *Evaluating Training Programs: The Four Levels*, 2nd Edition. San Francisco: Berrett-Koehler Publishers, 1998.

Phillips, Jack J. *Handbook of Training Evaluation and Measurement Methods*, 3rd Edition. Houston: Gulf Publishing, 1997.

Figure 6.7. Downloadable personalized action steps.

Data Collection Ideas to Pursue/Actions To Take When Preparing To Collect Data			
Action Item/Issue	*Responsible Person*	*Others Who Can Help*	*Date To Be Completed*

Phillips, Jack J. 1994, 1997. *Measuring The Return On Investment.* Vol 1 and Vol 2. Alexandria, VA: American Society for Training and Development.

Phillips, Jack J. "Return On Investment in Training and Performance Improvement Programs." Houston, TX: 1997, Gulf Publishing.

Phillips, Jack J. "Was It The Training?" *Training & Development*, Vol. 50, No. 3, March 1996, pp. 28-32.

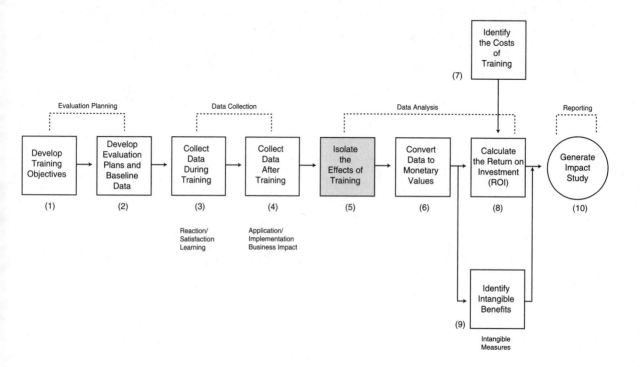

Evaluation Planning

Develop
Training
Objectives
(1)

Develop
Evaluation
Plans and
Baseline
Data
(2)

Data Collection

Collect
Data
During
Training
(3)

Reaction/
Satisfaction
Learning

Collect
Data
After
Training
(4)

Application/
Implementation
Business Impact

Data Analysis

Isolate
the
Effects of
Training
(5)

Convert
Data to
Monetary
Values
(6)

Identify
the Costs
of
Training
(7)

Calculate
the Return on
Investment
(ROI)
(8)

Identify
Intangible
Benefits
(9)

Intangible
Measures

Reporting

Generate
Impact
Study
(10)

7

Step 5. Isolate the Effects of Training

When business measures are influenced by actions taken within the organization or by external factors, executives want to know what causes the influence. If the actions or factors cause a favorable influence, such as an increase in sales or an improvement in quality, management would like to sustain the movement and/or replicate it in other parts of the organization. If the actions are having a negative effect, management would like to design and implement strategies to turn the situation around. The case illustration "First Bank" demonstrates this point.

CASE ILLUSTRATION: FIRST BANK

First Bank, a large commercial bank, had experienced a significant increase in consumer loan volume for the quarter. In an executive meeting, the chief executive officer asked the executive group why the volume had increased.

- The executive responsible for the consumer lending started the discussion by pointing out that his loan officers are more aggressive. "They have adopted an improved sales approach. They all have sales development plans in place. We are being more aggressive."

- The marketing executive added that she thought the increase was related to a new promotional program and an increase in advertising during the period. "We have had some very effective ads," she remarked.

- The chief financial officer thought it was the result of falling interest rates. He pointed out that interest rates fell by an average of 1% for the quarter and added, "each time interest rates fall, consumers will borrow more money."

- The executive responsible for mergers and acquisitions felt that the change was due to a reduction in competition. "Two bank branches were closed during this quarter, which had an impact on our market areas. This has driven those customers over to our branches." She added, "When you have more customers, you will have more loan volume."

- The human resources vice president spoke up and said that the consumer loan referral incentive plan had been slightly altered with an increase in the referral bonus to all employees who refer legitimate customers for consumer loans. This new bonus plan, in her opinion, had caused the increased consumer loan value. She concluded, "When you reward employees to bring in customers, they will bring them in . . . in greater numbers."

- The human resource development vice president said the consumer lending seminar caused the improvement. He indicated that it had been revised and is extremely effective, with appropriate strategies to increase customer prospects. He concluded, "When you have effective training and build skills in sales, you will increase loan volume."

Each of these responses probably has some validity. Each likely contributed. The only way to know is to utilize one or more techniques to isolate the effects.

Similarly, when a significant increase in performance appears after a training program has been conducted, the two events appear to be linked. While the change in performance may be linked to the training, other factors usually have also contributed to the improvement. The cause-and-effect relationship between training and performance can be very confusing and difficult to prove, but can be accomplished with an acceptable degree of accuracy.

Isolating the effects of your program is one of the most important and most challenging parts of the ROI Process. The credibility of your study, to a significant degree, hinges on how effectively you separate out the training's impacts from other influences in your organization's environment. This chapter will help you understand and select the best isolation technique to use in different circumstances. Ten useful strategies are explored for isolating the effects of training when collecting Level-3 and -4 data. These strategies have been utilized by many organizations to determine the amount of the improvement that is directly related to training.

IDENTIFYING OTHER FACTORS: A FIRST STEP

As a first step in the isolating process, all of the key factors that may have contributed to the performance improvement should be identified. This step communicates to interested parties that other factors may have influenced the results, underscoring that the program is not the sole source of improvement. Consequently, the credit for improvement is shared with several possible sources, an approach that is likely to gain the respect of management. Figure 7.1 illustrates how other factors can influence results.

There are many factors, events, and processes that can influence output variables even though the training is designed to focus directly on a specific improvement. Management attention is a tremendous influence; it has two parts. One part is the Hawthorn effect, in which improvements are often realized just because attention is focused on them. The fact that training is provided on improving customer ser-

Figure 7.1. Factors contributing to an improvement after a training program is conducted.

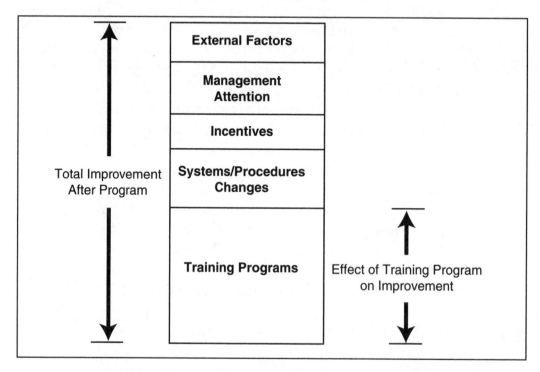

vice will often improve the service some regardless of the skills learned and applied from the program. The second part is the Pygmalion effect, in which participants often improve their performance because they are expected to improve. That is, if management expects participants to apply the skills taught and achieve results, they will often follow through and do that just because it is expected. This is in addition to what they actually learned.

There are several potential sources to identify major influencing variables:

- Clients. If the program is designed on request, the client may be able to identify factors that will influence the output variable. The client will usually be aware of other initiatives or programs that may impact the output.

- Training program participants. Training participants are usually aware of other influences that may have caused performance improvement. After all, it is the impact of their collective efforts that is being monitored and measured. In many situations, they have witnessed previous movements in the performance measures and can pinpoint the reasons for changes.

- Performance analysts, designers and developers. Training and Development designers and analysts are another source for identifying variables that have an impact on performance results. The needs analysis will usually uncover these influencing variables. Designers usually analyze these variables while addressing the issue of transfer of learning.

- Supervisors of participants. Immediate supervisors, in addition to the participants, are often in a position to see and understand the numerous variables that influence performance in work situations. They have performance discussions with their team members from time to time, and this can provide them with unique insights about performance issues.

- Middle and top management. Senior management may be able to identify other influences based on their experience and knowledge of the situation. Perhaps they have monitored, examined, and analyzed the variables previously. The authority position of these individuals often increases the credibility of the data.

Taking time to focus attention on variables that may have influenced performance brings additional accuracy and credibility to the process. It moves beyond the scenario where results are presented with no mention of other influences, a situation that often destroys the credibility of an impact report. It also provides a foundation for some of the strategies described in this chapter by identifying the variables that must be isolated to show the effects.

THE BEST STRATEGIES TO ISOLATE THE EFFECTS OF TRAINING

The ten isolating strategies are:

- Control groups
- Trend line analysis
- Forecasting
- Participant estimate
- Supervisor estimate
- Management estimate
- Customer input
- Expert estimate
- Subordinate input
- Other factors impact

These strategies are summarized below.

Control Groups. The most accurate isolation approach is the use of control groups in an experimental design process. This involves the use of an experimental group that is impacted by the program and a control group that is not. The composition of both groups should be as identical as possible and, if feasible, the selection of participants for each group should be on a random basis. When this is possible and both groups are subjected to the same environmental influences, the differences in the performance of the two groups can be attributed to the training program. An example of control group utilization for a customer service training program is explained below.

- Six sites are chosen for program evaluation.
- Each site has a control group and an experimental group randomly selected.
- The experimental group receives training; the control group does not.
- Performance is observed for both groups during the same time period.

- Evaluation data is collected for both groups at the same time.

Because a control group is an effective isolating technique, it should be considered as a strategy when a major ROI impact study is planned. The primary advantage of the control group process is accuracy, and thus should be used in those situations where it is important for the impact to be isolated to a high level of accuracy.

Control group arrangements appear in many settings. A Federal Express company ROI analysis used control groups. The study focused on 20 employees who went through an intense, redesigned two-week training program soon after being hired to drive company vans. Their performance was compared with a control group of 20 other new hires whose managers were told to do no more or less on-the-job training than they normally would. Performance was tracked for the two groups for 90 days in categories such as accidents, injuries, time-card errors, and domestic air-bill errors. The ten performance categories were assigned dollar values by experts from engineering, finance, and other groups. The program demonstrated that the performance of the highly trained employees was superior to that of the group which did not receive the upgraded training and also resulted in a 24% return on investment.

The control group process does have some inherent problems that may make it difficult to apply in practice.

- From a practical perspective, it is virtually impossible to have identical control and experimental groups.
- Dozens of factors can affect employee performance, some of them individual and others contextual. To tackle the issue on a practical basis, it is best to select three or four variables that will have the greatest influence on performance.
- The Hawthorne/Pygmalion effect can occur to some degree if the control group knows that its performance is being observed and measured.
- Contamination can develop when participants in the training initiative actually teach others who are in the control group.

Sometimes the reverse situation occurs when members of the control group will model the behavior from the experimental group. In either case, the experiment becomes contaminated as the influence of the program is passed on to the control group. Contamination can be minimized by ensuring that control groups and experimental groups are at different locations. When isolation is not possible, it can be helpful to explain to both groups that one group will receive training now and another will participate in the training at a later date. Appeal to the sense of responsibility of those being trained and ask them not to share the information with others.

- If the groups are in different locations, they may have different environmental influences. Careful selection of the groups can help prevent this problem from occurring.

- Using control groups may appear to be too research oriented for training and development and sales management.

  - Management may not want to take the time to experiment before proceeding with a program or they may not want to withhold involvement in the training initiative from any group just to measure the impact of an experimental program. In part, this may be due to concerns of fairness or legality regarding incentive compensation for the control groups.

  - This concern can be offset by implementing the training initiative with pilot participants as the experimental group and select nonparticipants to serve as the control group. Under this arrangement, the control group is not informed of its control group status.

Trend Line Analysis. Another useful technique for approximating impact is trend line analysis. If historical performance data are available, a trend line can be drawn, using the previous performance data as a base and extending the trend into the future. After a training program is conducted, actual performance is plotted and compared to the trend line. Any improvement in actual performance over what the trend line predicted can then be reasonably attributed to the training. While this is not an exact process, it provides a reasonable estimation of the program's impact.

The primary advantage of this approach is that it is simple, inexpensive, and takes very little effort.

Figure 7.2 includes an example of this trend line analysis taken from a shipping department of a large distribution company. The percent of schedule shipped reflects the level of actual shipments compared to scheduled shipments. Data are presented before and after a team training program conducted in July. As shown, there was an upward trend on the data prior to the training program.

Although the program apparently had a dramatic effect on shipment productivity, the trend line shows that the improvement would have continued anyway, based on the trend that had been previously

Figure 7.2. Example of trend line analysis.

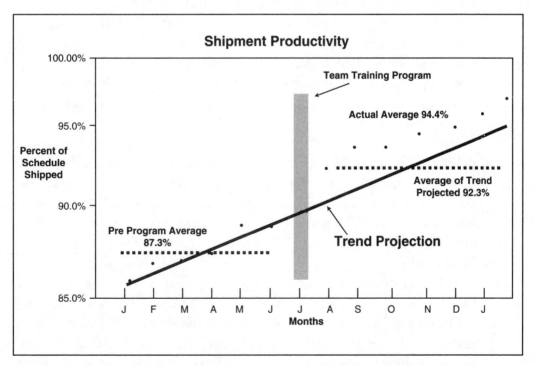

Reproduced from *Handbook of Training Evaluation and Measurement Methods,* Gulf Publishing, 1997, Jack J. Phillips, author.

established. It is tempting to measure the improvement by comparing the average six-months shipments prior to the program (87.3%) to the average six months after the program (94.4%), yielding a 6.9% difference. However, a more accurate comparison is to use the six-month average, after the program, compared to the trend line value at the same point of the trend line (92.3%). In this example, the difference is 2.1%. Using this more modest measure increases the accuracy and credibility of the process.

Forecasting Methods. A more analytical approach to trend line analysis is to use forecasting methods to predict a change in performance variables. This approach represents a mathematical interpretation of the trend line analysis when other variables enter the situation. A linear model, in the form of $y = ax + b$, is appropriate when only one other variable influences the output performance and that relationship is characterized by a straight line. Instead of drawing the straight line, a linear equation is developed from which to calculate a value of the anticipated performance improvement.

An example will help explain the application of this process. A large retail store chain implemented a sales training program for sales associates. The three-day program was designed to enhance sales skills and prospecting techniques. The application of the skills should increase the sales volume for each sales associate. An important measure of the program's success is the sales per employee six months after the program when compared to the same measure prior to the program. The average daily sales per employee prior to training, using a one-month average, was $1100. Six months after the program, the average daily sales per employee was $1500. Two related questions must be answered: Is the difference in these two values attributable to the training? Did other factors influence the actual sales level?

After reviewing potential influencing factors with several executives, only one factor, the level of advertising, appeared to have changed significantly during the period under consideration. When reviewing the previous sales-per-employee data and the level of advertising, a direct relationship appeared to exist. As expected, when advertising expenditures were increased, the sales per employee increased proportionately.

Using the historical values to develop a simple linear model yielded the following relationship: $y = 140 + 40x$, where y is the daily sales per employee and x is the level of advertising expenditures per week (divided by 1000). The development of this equation is a process of deriving a mathematical relationship between two variables using the method of least squares. This is a routine option on some calculators and is included in many software packages.

The level of weekly advertising expenditures in the month preceding training was $24,000, and the level of expenditures in the sixth month after training was $30,000. Assuming that the other factors that might influence sales were insignificant, as concluded by the store executives, the impact of the training program is determined by plugging in the new advertising expenditure amount, 30, for x and calculating the daily sales, which yields $1340.

Thus, the new sales level caused by the increase in advertising is $1340, as shown in Figure 7.3. Since the new actual value is $1500, then $160 (1500 – 1340) must be attributed to the training program.

Participant Estimation of Impact. An easily implemented method to isolate a training program's impact is to obtain information directly from program participants. The effectiveness of this approach rests on the assumption that participants are capable of determining or estimating how much of a performance improvement is related to the program. Because their actions have produced the improvement, participants may have very accurate input on the issue. They should know how much of the change was caused by applying what they learned. Although an estimate, this value will usually have considerable credibility with management because participants are at the center of the change or improvement. Participant estimation is obtained by asking a series of questions after describing the improvement.

- What percent of this improvement can be attributed to the application of skills/techniques/knowledge gained in the training program?

Figure 7.3. Forecasting daily sales based on advertising.

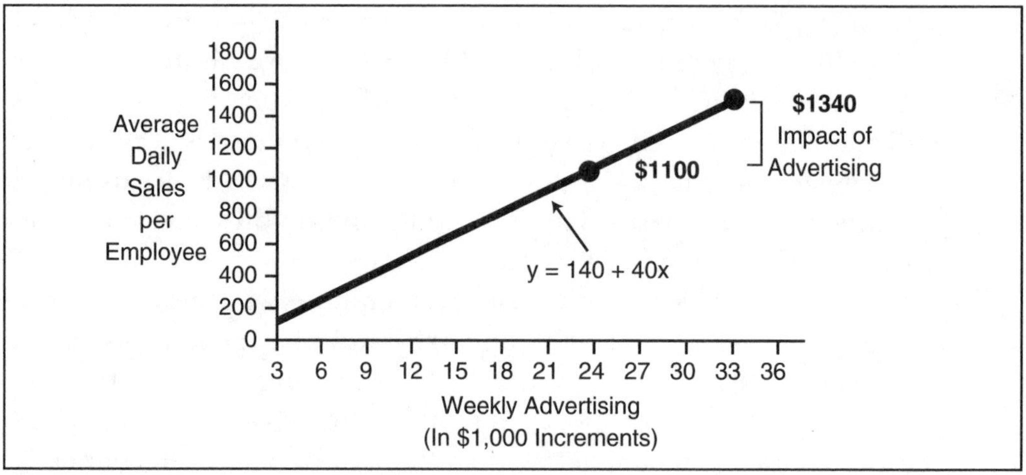

Reproduced from *Handbook of Training Evaluation and Measurement Methods,* Gulf Publishing, 1997, Jack J. Phillips, author.

- What is the basis for this estimation?
- What confidence do you have in this estimate, expressed as a percentage?
- What other factors contributed to this improvement in performance?
- What other individuals or groups could estimate this percentage or determine the amount?

Although this is an estimate, this approach does have considerable accuracy and credibility. Five adjustments are effectively utilized with this approach to reflect a conservative approach:

1. The individuals who do not respond to the questionnaire or provide usable data on the questionnaire are assumed to have no improvements. This is probably an overstatement, since some individuals will have improvements but not report them.

2. Extreme data and unrealistic claims are omitted from the analysis, although they may be included in the report of intangible benefits.

3. Since only annualized values are used, it is assumed that there are no benefits from the program after the first year of implementation. In reality, some training programs should be expected to add value for several years after being conducted.
4. The confidence level, expressed as a percentage, is multiplied by the improvement value to reduce the amount of the improvement by the potential error.
5. The improvement amount is adjusted by the amount directly related to the program, expressed as a percentage.

As an added enhancement to this method, management may be asked to approve the amounts that have been estimated by participants. Table 7.1 is an example of participant estimates.

Each participant's estimate is adjusted based on the principles mentioned above. For example, the value provided by participant "A" is adjusted as follows:

A $135,000	Faster production of product drawings. 40% 80% $43,200
Annual improvement $135,000 due to faster production	
Confidence level	$135,000 x 0.40 confidence level = $54,000
Factor for training impact	$54,000 x 0.80 change caused by program = $43,200
Total annualized impact	$43,200

The case of the National Bank shows one way that participants can estimate the impact of various factors. National Bank uses sales training coupled with incentives and management reinforcement to ensure that ambitious sales goals are met. Salespeople aggressively pursue new customers and cross-sell existing customers in a variety of product lines. The following factors were initially identified by branch managers as having a significant influence on sales output:

1. The sales training program
2. Incentive systems
3. Management reinforcement/management emphasis
4. Market fluctuations

Table 7.1. Sample of participants' estimates from a leadership program for managers.

PARTICIPANT	ANNUAL IMPROVEMENT VALUE	BASIS FOR VALUE	CONFIDENCE FACTOR	ISOLATION FACTOR	ADJUSTED VALUE
A	$135,000	Faster production of product drawings.	40%	80%	$ 43,200
B	$112,100	Increase in output, with man-hour savings due to better management of backlog.	98%	40%	$ 43,943
C	$9,240	Reduction in overtime.	100%	50%	$ 4,620
D	$3,600	Improved efficiency in export order process.	75%	100%	$ 2,700
E	$100,000	Just-in-time delivery scheduling resulted in reduction in inventory requirements.	85%	20%	$17,000
F	$18,000	Reduced site management time for rework by one week during warranty liability period.	80%	80%	$11,520
G	$10,600	Saved planning time and reallocated time to productive tasks.	75%	10%	$ 795
Total Annual Monetary Benefits					$123,778

The bank kept track of the sales of each product line, such as new loans, new checking accounts, and new credit card accounts. The branch managers in the target area were contacted by questionnaire six months after the training and asked to estimate the percent of improvement that could be attributed to each of the factors above. All branch employees provided input in a meeting that was facilitated by the manager. In the carefully organized meeting, the branch manager:

- Described the task.
- Explained why the information was needed and how it will be used.
- Had employees discuss the linkage between each factor and the specific output measure.
- Provided employees with any additional information needed to estimate the contribution of each factor.
- Asked employees to identify any other factors that may have contributed to the increase.
- Obtained the actual estimate of the contribution of each factor. The total of all factors had to be 100%.
- Obtained the confidence level from each employee for the estimate for each factor (100% = certainty; 0% = no confidence). The values are averaged for each factor.

Table 7.2 shows the information that was collected from one branch, for one of the product line business measures. The increase in accounts during the reporting period is 175.

The number of new credit card accounts per month that can be attributed to the sales program can be calculated from the estimates by multiplying the percent for the program factor by the total amount of improvement for each product line. This shows the impact of the program for that product line. The value is then adjusted by the confidence percentage.

The monthly increase in credit card accounts was 175. The number of these new accounts attributable to the training was 46 per

Table 7.2. Monthly increase in credit card accounts.

CONTRIBUTING FACTORS	CONFIDENCE IMPACT ON RESULTS (PERCENT)	AVERAGE CONFIDENCE LEVEL (PERCENT)
Sales training program	32%	83%
Incentive Systems	41%	87%
Management reinforcement/ management emphasis	14%	62%
Market fluctuations	11%	75%
Other _____	2%	91%
	100%	

month. This is determined by multiplying 175 x 0.32 isolation factor, which yields 56. Then multiply 56 by the 0.83 confidence level and the answer is 46.48 or 46 credit card accounts.

Supervisor Estimates of the Impact of Training. In lieu of, or in addition to, participant estimates, the participants' supervisor may be asked to provide input as to the extent of the training program's role in producing a performance improvement. In some settings, supervisors of participants may be more familiar with the other factors that influence performance. Consequently, they may be better equipped to provide estimates of impact. The recommended questions to ask supervisors are essentially the same as those in the participants' questionnaire. Supervisor estimates should be analyzed in the same manner as participant estimates. The advantages and disadvantages of this approach are similar to those for participant estimation.

Management's Estimate of Training's Impact. In some cases, upper management may estimate the percent of improvement that should be attributed to the training program. Although this process is very subjective, the input is received from the individuals who often provide or approve funding for the program. Sometimes their level of comfort with the process is the most important consideration.

Customer Input. Another helpful approach in some narrowly focused situations is to solicit input on the impact of a training program directly from customers. In these situations, customers are asked why they chose a particular product or service or to explain how their reaction to the product or service has been influenced by individuals' applying specific skills and abilities (those targeted by the training program). This strategy focuses directly on behaviors the program is designed to improve.

Expert Estimation of Impact. External or internal experts can sometimes provide an estimate on the portion of results that can be attributed to a program. When using this strategy, experts must be carefully selected based on their knowledge of the process and situation. For example, an expert in quality might be able to provide estimates of how much change in a quality measure can be attributed to training and how much can be attributed to other factors.

This approach does have disadvantages. It can be inaccurate unless the program and setting on which the estimate is made is very similar to the program in question. Also, this approach may lose credibility when the estimates come from external sources because this may not necessarily involve those who are close to the process.

This process has an advantage in that its credibility often reflects the reputation of the expert or independent consultant. It is a quick source of input from a reputable expert or independent consultant. Sometimes top management will place more confidence in external experts than in their own internal staff. However, internal experts are often viewed as credible.

Subordinate Input on Impact. In some situations, the subordinates of the participants being trained can provide input concerning the extent of a program's impact. Although they will not usually be able to estimate how much of an improvement can be attributed to the program, they can provide input in terms of what other factors might have contributed to the improvement. This approach is appropriate where managers are being trained to implement work unit changes or develop new skills for use with employees. Improvements are realized through the utilization of the skills. The supervisor's employees provide input about changes that have occurred since the training was conducted. They help determine the extent to which other factors have changed in addition to supervisor behavior.

Subordinate input is usually obtained through surveys or interviews. When the survey results show significant changes in supervisor behavior after training and no significant change in the general work climate, the improvement in work performance, therefore, must be attributed to the changes in supervisor behavior, since other factors remained constant.

This approach has some disadvantages. Data from subordinates is subjective and may be questionable because of the possibility for biased input. Also, in some cases the subordinates may have difficulty in determining changes in the work climate. However, in some cases, subordinates are aware of the factors that caused changes in their work unit and can provide input about the magnitude or quantity of these changes. When combined with other methods to isolate impact, this process has increased credibility.

Calculating the Impact of Other Factors. Although not appropriate in all cases, there are some situations in which it may be feasible to calculate the impact of factors other than the training program that have influenced the improvement and then conclude that the program can be credited with the remaining impact. In this approach, the program takes credit for improvement that cannot be attributed to other factors.

An example will help explain this approach. In a consumer lending program for a large commercial bank, a significant increase in consumer loan volume was generated after a training program was conducted for consumer loan officers. Part of the increase in volume was attributed to training and the remaining was due to the influence of other factors operating during the same time period. Two other factors were identified by the evaluator: a loan officer's production improved with time, and falling interest rates caused an increase in the volume of consumer loans.

In regard to the first factor, as loan officers make loans, their confidence improves. They use consumer lending policy manuals and gain knowledge and expertise through trial and error. The amount of this factor was estimated by using input from several internal experts in the marketing department.

For the second factor, industry sources were utilized to estimate the relationship between increased consumer loan volume and falling interest rates. These two estimates together accounted for a certain percent of increased consumer loan volume. The remaining improvement was attributed to the training program.

This method is appropriate when the other factors are easily identified and the appropriate mechanisms are in place to calculate their impact on the improvement. In some cases it is just as difficult to estimate the impact of other factors as it is to estimate the impact of the training program, leaving this approach less advantageous. This process can be very credible if the method used to isolate the impact of other factors is credible.

DECIDING WHICH STRATEGIES TO USE

One key factor in the success of your evaluation will be selecting the most appropriate strategy with which to isolate the effects of training. Here are the essential questions to ask:

- Is a control group arrangement feasible?
- Are historical data available?
- Have mathematical relationships been developed between input factors and output measures?
- Are participants capable and willing to estimate the effect of training on output measures?
- Are the supervisors of participants capable and willing to estimate the effect of training on output measures?
- Is senior management capable and willing to estimate the effect of training on output measures?

- Can vendors or experts provide realistic estimates of the effect of training?
- Are subordinates of participants capable of providing input on potential changes in the work climate?
- Can the effect of nontraining factors on output measures be determined?
- Can internal or external customers provide information on the influence of training on customer decisions and customer satisfaction?
- Are there any constraints on the use of data collection instruments?

Several other factors should also be considered:

- Feasibility of the strategy
- Accuracy provided with the strategy
- Credibility of the strategy with the target audience
- Specific cost to implement the strategy
- The amount of disruption in normal work activities as the strategy is implemented
- Participant, staff, and management time required by the particular strategy

Multiple strategies or multiple sources for data input should be considered, as two sources are usually better than one. When multiple sources are utilized, a conservative method is recommended to combine the inputs. A conservative approach builds acceptance.

It is not unusual for training ROIs to be extremely large. Even when a portion of the improvement is allocated to other factors, the numbers are still impressive in many situations. The audience should understand that although every effort was made to isolate the impact, it is still a figure that is not precise and may contain error. It represents the best estimate of the impact given the conditions, time, and resources that the organization is willing to commit to the process. Chances are it is more accurate than other types of analysis being performed in the organization.

When using estimates it is important to remember that you may be challenged. The following is generally true about estimates.

- Estimations are considered appropriate by experts
- Estimations are considered acceptable to management
- The estimator must have credibility
- The estimator should have authority from position or expertise

The credibility of outcome data is often in the eyes of the beholder. Most people will judge data on the following criteria:

CRITERIA	ISSUE IN QUESTION
1. Reputation of the source of the data	What is the capability or reliability of the source of the data?
2. Reputation of the source of the study?	What is the credibility of those involved in administering the study?
3. Motives of the researchers.	What interest do the researchers have in the outcome of the study?
4. Methodology of the study.	Is the methodology systematic, conservative, and is it thorough?
5. Assumptions made in the analysis.	Are the assumptions stated, are they thorough, and given the situation, are they reasonable?
6. Realism of the outcome data.	How realistic is the data; is it too complicated to draw reasonable conclusions; is it relevant to the organization's issues and needs?
7. Type of data.	Is it hard data and is it objective in nature or is it soft data and is it subjective in nature?
8. Scope of analysis.	Is the scope of the study narrow and therefore easier to see cause and effect, or is it broad in scope with many variables and influences?

Whatever method(s) you use to isolate the effects of training, it is important that you remain conservative in your calculations.

The worksheet in Figure 7.4 can be used to develop an isolation question to be included in a questionnaire to participants. The same approach can be used for other stakeholders. Figure 7.5 includes an example of isolating the effects using participant estimates.

Figure 7.4. Downloadable worksheet—Isolating effects using estimates.

Several factors often contribute to performance improvement. In addition to the
_____training, other potential factors are identified below. Look at the factors and indicate what percentage of your overall performance improvement during the past six months you attribute to each of the factors. If you feel that some of the factors had no influence, do not assign them a percentage. *The total of all selected items must equal 100 percent.*

Please select the items that you feel are appropriate by writing in your estimated percentages.	*Write in the percentage attributed to appropriate items.*
Factors	
A)	%
B)	%
C)	%
D)	%
E)	%
F)	%
G)	%
I) Other (please specify):	
Total of all selected items must = 100 percent	**Total 100%**

Copyright McGraw-Hill 2002. To customize this handout for your audience, download it from (www.books.mcgraw-hill.com/training/download). The document can then be opened, edited, and printed using Microsoft Word or other word-processing software.

Figure 7.5. Example—Isolating effects using estimates.

Several factors often contribute to performance improvement. In addition to the _____training, other potential factors are identified below. Look at the factors and indicate what percentage of your overall performance improvement during the past six months you attribute to each of the factors. If you feel that some of the factors had no influence, do not assign them a percentage. *The total of all selected items must equal 100 percent.*

Please select the items that you feel are appropriate by writing in your estimated percentages.	*Write in the percentage attributed to appropriate items.*
Factors	
A) System changes	15%
B) Training project	60%
C) Coaching by supervisor	8%
D) Compensation changes	17%
E) Other (please specify):	%
Total of all selected items must = 100 percent	**Total 100%**

FURTHER READING

Phillips, Jack J. *Handbook of Training Evaluation and Measurement Methods,* 3rd Edition. Houston: Gulf Publishing, 1997.

Phillips, Jack J. "Return On Investment in Training and Performance Improvement Programs." Houston, TX: 1997, Gulf Publishing.

Phillips, Jack J. "Was It The Training?" *Training & Development,* Vol. 50, No. 3, March 1996, pp. 28-32.

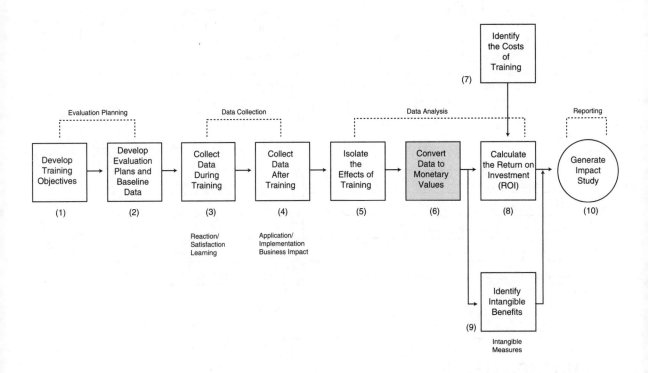

Identify
the Costs
of
Training

(7)

Evaluation Planning

Data Collection

Data Analysis

Reporting

| Develop Training Objectives | Develop Evaluation Plans and Baseline Data | Collect Data During Training | Collect Data After Training | Isolate the Effects of Training | Convert Data to Monetary Values | Calculate the Return on Investment (ROI) | Generate Impact Study |

(1)   (2)   (3)   (4)   (5)   (6)   (8)   (10)

Reaction/
Satisfaction
Learning

Application/
Implementation
Business Impact

Identify
Intangible
Benefits

(9)

Intangible
Measures

8

Step 6. Convert Data to Monetary Values

Most training-impact evaluations stop with a tabulation of business results, detailing improvements such as quality enhancements, reduced absenteeism, and improved customer satisfaction. While these results are important, it is more meaningful to compare the monetary value of the results to the cost of the training, so the true value of the training to the organization can be assessed. Converting data to monetary values is the first phase in putting training initiatives on the same level as other investments that organizations make. The key is to express outcomes in the language of business: money.

This chapter introduces a systematic approach to transforming outcome data into money terms. It tells how to categorize data accurately as "hard" or "soft" and how to select the best strategies for converting data to monetary values for specific programs.

SORTING OUT HARD AND SOFT DATA

It is essential to the ROI Process to divide data into hard and soft categories.

Hard data are the traditional measures of organizational performance—very objective, easy to measure, and easy to convert to monetary values. Hard data are common measures, have high credibility with management, and are available in every type of organization. These measures represent the output, quality, cost, and time of work-related processes. The characteristics of hard data are:

- Objectively based
- Easy to measure and quantify
- Relatively easy to assign monetary values to
- Common measures of organizational performance
- Very credible with management

Almost every unit will have hard-data performance measures. For example, consider a state government office approving applications for new drivers licenses. Among its overall performance measures will be the number of applications processed (output), cost per application processed (cost), the number of errors made in processing applications (quality), and the time it takes to process and approve an application (time). Ideally, training interventions for the people involved in doing the work in this unit would be linked to one or more hard-data measures. Table 8.1 contains examples of hard data.

Because many training interventions are designed to impact "soft" variables, the category of soft data is also needed in an evaluation. Soft data are sometimes subjective, may be difficult to measure, and are almost always difficult to convert to monetary values. When compared to hard data, soft data are usually less credible as measures of performance.

Soft data can be grouped into several categories. Measures such as employee turnover, absenteeism, and grievances are considered soft data, not because they are difficult to measure but because it is difficult to accurately convert them to monetary values. The characteristics of soft data are:

- Subjectively based in many cases
- Difficult to measure and quantify, directly
- Difficult to assign monetary values to
- Less credible as measures of performance
- Usually behaviorally oriented

Table 8.2 includes examples of soft data.

Table 8.1. Examples of hard data.

OUTPUT	TIME
Units produced	Equipment downtime
Tons manufactured	Overtime
Items assembled	On-time shipments
Money collected	Time to project completion
Items sold	Processing time
Forms processed	Supervisory time
Loans approved	Break in time for new employees
Inventory turnover	Learning time
Patients visited	Meeting schedules
Applications processed	Repair time
Students graduated	Efficiency
Tasks completed	Work stoppages
Output per hour	Order response
Productivity	Late reporting
Work backlog	Lost-time days
Incentive bonus	**QUALITY**
Shipments	Scrap
New accounts generated	Waste
COSTS	Rejects
Budget variances	Error rates
Unit costs	Rework
Cost by account	Shortages
Variable costs	Product defects
Fixed costs	Deviation from standard
Overhead cost	Product failures
Operating costs	Inventory adjustments
Number of cost reductions	Time-card corrections
Project cost savings	Percent of tasks completed
Accident costs	properly
Program costs	Number of accidents
Sales expense	

Table 8.2. Examples of soft data.

WORK HABITS Absenteeism Tardiness Visits to the dispensary First aid treatments Violations of safety rules Number of communication break-downs Excessive breaks Follow-Up **WORK CLIMATE/SATISFACTION** Number of grievances Number of discrimination charges Employee complaints Job satisfaction Employee turnover Litigation Organizational commitment Employee loyalty Increased confidence **CUSTOMER SERVICE** Customer complaints Customer satisfaction Customer dissatisfaction	Customer impressions Customer loyalty Customer retention Customer value Lost customers **EMPLOYEE DEVELOPMENT/ADVANCEMENT** Number of promotions Number of pay increases Number of learning programs attended Requests for transfer Performance appraisal ratings Increases in job effectiveness **INITIATIVE/INOVATION** Implementation of new ideas Successful completion of projects Number of suggestions implemented Setting goals and objectives New products and services developed New patents and copyrights

THE BEST STRATEGIES FOR CONVERTING DATA TO MONETARY VALUES

Ten strategies are available to convert data to monetary values. Some strategies are appropriate for a specific type or category of data, while other strategies can be used with virtually any type of data. The challenge is to select the strategy that best matches the type of data and situation. Figure 8.1 is a list of the ten strategies for converting data to monetary values. Following the figure, each strategy is presented, beginning with the most credible approach.

1. Converting output to contribution
2. Converting the cost of quality
3. Converting employee time
4. Using historical costs
5. Using internal and external experts
6. Using data from external databases
7. Using participants' estimates

Figure 8.1. Best strategies for converting data.

1. Converting output to contribution
2. Converting the cost of quality
3. Converting employee time
4. Using historical costs
5. Using internal and external experts
6. Using internal and external databases
7. Using participants' estimates
8. Linking with other measures
9. Using supervisors' and managers' estimates
10. Using training staff estimates

Converting output data to contribution

When a training program has produced a change in performance (output), the value of the increased output can usually be determined from accounting or operating records in the organization. For organizations operating on a profit basis, this value is usually the marginal profit contribution of an additional unit of production or unit of service. For example, a production team at a major appliance manufacturer is able to boost production of small refrigerators as the result of a series of highly focused training programs. The unit of improvement is the profit margin of one refrigerator. In not-for-profit organizations, this value is usually reflected in the savings accumulated when an additional unit of output is realized for the same input requirements. For example, in a visa section of a government office, an additional visa application is processed at no additional cost. Thus, an increase in output translates into a cost savings equal to the unit cost of processing a visa.

The formulas and calculations used to measure this contribution depend on the organization and its records. Most organizations have this type of data readily available for performance monitoring and goal setting.

In one case involving a commercial bank, a sales seminar for consumer loan officers was conducted that resulted in additional consumer loan volume (output). To measure the return on investment in the training program, it was necessary to calculate the value (profit contribution) of one additional consumer loan. From the bank's records, it was a relatively easy item to calculate and had been developed on a regular basis. As shown in Table 8.3, several components went into this calculation.

The first step was to determine the yield, which was available from bank records. Next, the average spread between the cost of funds and the yield received on the loan was calculated. For example, the bank could obtain funds from depositors at 5.5 percent on average, including the cost of operating the branches. The direct costs of making

Table 8.3. Loan profitability analysis.

PROFIT COMPONENT	UNIT VALUE
Average loan size	$15,500
Average loan yield	9.75%
Average cost of funds (including branch costs)	5.50%
Direct costs for consumer lending	0.82%
Corporate overhead	1.61%
Net Profit Per Loan	**1.82%**

the loan, such as the salaries of the employees directly involved in consumer lending and the costs of advertising for consumer loans, had to be subtracted from this difference. Historically, these direct costs amounted to 0.82 percent of the loan value. To cover overhead costs for all the other corporate functions, an additional 1.61 percent was subtracted from the value. The remaining 1.82 percent of the average loan value represented the profit margin of a loan.

Calculating the cost of quality

The cost of quality is an important measure in most firms. Since many training programs are designed to improve quality, a value must be placed on the improvement in certain quality measures. If quality is measured with a defect rate, the value of the improvement is the cost to repair or replace the product. The most obvious cost of poor quality is the scrap or waste generated by mistakes. Many organizations use a standard value when determining the cost of poor quality. For example, one company that processes and manufactures gelatin that is used to encapsulate medication uses the cost of raw materials (gelatin) as the standard cost for gelatin waste in the plant. As an example, if the manufacturing operation wastes 9000 kilograms monthly at a raw materials cost of $4.20 per kilogram, this converts to a monthly cost of $37,800 (9000 x $4.20 = $37,800). If a training program to reduce waste had an impact of cutting the waste in half to 4500 kilograms per month, the savings would be $18,900 monthly or $226,800 annually. This savings would be reduced by the cost of the training program.

Perhaps the costliest element of poor quality is the dissatisfaction of customers and clients when mistakes are made. Customer dissatisfaction is difficult to quantify, and it may be impossible to arrive at a monetary value using direct methods. Usually the judgment and expertise of sales, marketing, or quality managers are the best sources to measure the impact of dissatisfaction.

Converting employee time

Reduction in time spent on tasks by employees is a common performance-improvement objective. Time management training may be designed to help individual employees save time in performing tasks or to enable a team to perform tasks in a shorter time frame or with fewer people The value of the time saved is an important measure of the program's success, and the conversion to monetary values is a relatively easy process.

The most obvious savings are from reduction in labor costs for performing work. The monetary savings are the hours saved times the labor cost per hour. For example, after attending a time management training program, participants estimated an average time savings of 74 minutes per day, worth $31.25 per day or $7500 per year. This time savings was based on the average salary plus benefits for the typical participant.

In addition to the labor cost per hour, other benefits can result from time savings. These include improved service, avoiding penalties for late projects, and creating additional opportunities for profit.

A caveat: time savings is realized only when the time saved is used in a productive way. One means of addressing this is detailed in Figure 6.3 in Chapter 6. An example of capturing time savings and converting it to a monetary value is illustrated in Figure 6.5.

Using historical costs

Sometimes the value of a measurement will be contained in historical records that reflect the cost (or value) of a unit of improvement. This conversion strategy involves identifying the appropriate records and tabulating the actual cost components for the item in question. For example, a training program was implemented to improve safety performance for a large construction firm. Examining the records of the company using one year of data enabled the training staff to calculate the average cost for an accident and is illustrated in Table 8.4.

Using the figures from Table 8.4, if a training program could effect a 50% reduction in accidents (27 accidents), this would cut accident costs by $686,448 (27 accidents x $25,424 = $686,448). Historical data are usually available for most hard measures. Unfortunately, historical records are not usually available for soft measures; other strategies, explained in this chapter, must be employed to convert soft data to monetary values.

Using internal and external experts

When faced with converting soft data when historical records are not available, it might be feasible to elicit input from experts in the particular processes. Individuals within the organization who are very knowledgeable about the situation and also have earned the respect of management often are the best prospects for expert input. These experts are asked to provide the cost (or value) of one unit of improvement. These experts must understand the processes and be willing to provide estimates along with the assumptions used in arriving at the estimates. When requesting input from these individuals, it

Table 8.4. Costs of an accident.

THE COST OF AN ACCIDENT	
Direct medical costs related to accidents	$114,390.00
Worker compensation payments	327,430.00
Insurance premiums	120,750.00
Legal expenses	75,600.00
Total operating budget for safety and health department (minus the above values) including salaries and benefits of safety staff	455,280.00
Management and supervisory time devoted to accident prevention and investigation	105,000.00
Safety training costs not included in above operating budget	31,000.00
Safety awareness materials	31,000.00
Lost productivity for safety training, safety meetings, accident investigation, and replacement staff	95,000.00
Total	**$1,347,450.00**
Total number of accidents	**53**
Cost per accident	**$25,423.60**

Return on Investment in Training and Performance Improvement Programs, Jack J. Phillips, Ph.D., Butterworth-Heinemann, 1997.

is best to explain the full scope of what is needed, providing as many specifics as possible. Most experts have their own methodology for developing computations of value.

When internal experts are not available, external experts are sought. External experts must be selected based on their experience with the unit of measure. Fortunately, many experts are available who work directly with important measures such as employee attitudes, customer satisfaction, turnover, absenteeism, and grievances. They are often willing to provide estimates of the cost (or value) of these items. Because the credibility of the value is directly related to his or her reputation, the credibility and reputation of the expert is crucial.

Using values from external databases

For some soft data items, it may be appropriate to use estimates of the cost (or value) of one unit based on the research of others. This strategy taps external databases that contain studies and research projects focusing on the cost of data items. Fortunately many databases are available that report studies of costs of a variety of data items related to important measures. Data are available on the costs of turnover, absenteeism, grievances, accidents, and customer satisfaction. The difficulty lies in finding a database relevant to the program or population under evaluation. Table 8.5 shows an example of research conducted on the cost of turnover. An evaluator searching for the value of turnover for chemical engineers could use the database to determine the value and therefore the savings resulting from a training program that reduced turnover. For example, at an annual salary of $70,000 if the turnover were reduced from 30 chemical engineers per year to 10 per year, the gain of 20 would result in an annual savings of $140,000 per unit of turnover ($70,000 annual salary x 200% = $140,000 value for one turnover). This savings is calculated by using the conservative cost range of 200% from

Table 8.5. Turnover costs.

Job Type/Category	Turnover Cost Ranges as a Percent of Annual Wages/Salary
Entry level—hourly, nonskilled (e.g., fast food worker)	30–50%
Service/production workers—hourly (e.g., courier)	40–70%
Skilled hourly (e.g., machinist)	75–100%
Clerical/administrative (e.g., scheduler)	50–80%
Professional (e.g., sales representative, nurse, accountant)	75–125%
Technical (e.g., computer technician)	100–150%
Engineers (e.g., chemical engineer)	200–300%
Specialists (e.g., computer software designer)	200–400%
Supervisors/team leaders (e.g., section supervisor)	100–150%
Middle managers (e.g., department manager)	125–200%

Notes:
1. Percentages are rounded to reflect the general range of costs from studies.
2. Costs are fully loaded to include all the costs of replacing an employee and bringing him/her to the level of productivity and efficiency of the former employee. The turnover included in studies is usually unexpected and unwanted. The following costs categories are usually included:

Exit cost of previous employee	Lost productivity
Recruiting cost	Quality problems
Employee cost	Customer dissatisfaction
Orientation cost	Loss of expertise/knowledge
Training cost	Supervisor's time for turnover
Wages and salaries while training	Temporary replacement costs

3. Turnover costs are usually calculated when excessive turnover is an issue and turnover costs are high. The actual cost of turnover for a specific job in an organization may vary considerably. The above ranges are intended to reflect what has been generally reported in the literature when turnover costs are analyzed.

Sources of data
The sources of data for these studies follow 3 general categories:
1. Industry and trade magazines have reported the cost of turnover for a specific job within an industry.
2. Publications in general management (academic and practitioner), human resources management, human resources development training, and performance improvement often reflect ROI cost studies because of the importance of turnover to senior managers and human resources managers.
3. Independent studies have been conducted by organizations and not reported in the literature. Some of these studies have been provided privately to the authors.

the table. The total annual savings would be $2,800,000. ($140,000 per unit x annual improvement of 20 = $2,800,000).

Linking with other measures

When standard values, records, experts, and external studies are not available, a feasible approach might be developing a relationship between the measure in question and some other measure that easily can be converted to a monetary value. This involves identifying existing relationships, if possible, that show a strong correlation between one measure and another with a standard value.

For example, a classical relationship exists between increasing job satisfaction and employee turnover. In a program designed to improve job satisfaction, a value is needed for changes in the job satisfaction index. A predetermined relationship showing the correlation between improvements in job satisfaction and reductions in turnover can link the changes directly to turnover. Using standard data or external studies, the cost of turnover can easily be developed as described earlier. Thus, a change in job satisfaction is converted to a monetary value or, at least, an approximate value. It is not always exact because of the potential for error and other factors, but the estimate is sufficient for converting the data to monetary values.

In some situations, a chain of relationships may be established to show the connection between two or more variables. In this approach, a measure that may be difficult to convert to a monetary value is linked to other measures that, in turn, are linked to measures that can be expressed as values.

Using estimates from participants

In some situations, the value of a soft data improvement can be estimated by program participants. This strategy is appropriate where participants are capable of providing estimates of the cost (or value) of the unit of measure improved by applying the skills learned in the program. While this process makes some participants uneasy because they may be unable to accurately estimate the values, other participants are willing and capable of providing values. When using this approach, clear instructions should be provided to participants, along with examples of the type of information needed. The advantage of this approach is that the individuals closest to the improvement are often capable of providing the most reliable estimates of the value of the improvement.

Using estimates from supervisors and managers

In some situations, participants may be incapable of placing a value on the improvement. Their work may be so far removed from the output of the process that they cannot reliably provide estimates. In these cases, the team leaders, supervisors, or managers of participants may be asked to provide a value for a unit of improvement linked to the training program.

In some situations, senior managers who are interested in the process or program are asked to place a value on the improvement. This approach is used in situations where it is very difficult to calculate the value or other sources of estimation are unavailable or not reliable.

Using training staff estimates

The final strategy for converting data to monetary values is to use training staff estimates. Using all the available information, the staff members most familiar with the situation provide estimates of the value.

DECIDING WHICH STRATEGY TO USE TO CONVERT DATA

With so many strategies available, the challenge is to select one or more strategies appropriate to the situation. The following guidelines can help determine the proper selection.

Use the strategy appropriate for the type of data. Some strategies are designed specifically for hard data, while others are more appropriate for soft data. Consequently, the type of data will often dictate the strategy. Hard data, while always preferred, are not always available. Soft data are often required and must be addressed with the appropriate strategies.

Move from most-accurate to least-accurate strategies. The ten strategies are presented in the order of accuracy and credibility, beginning with the most credible. Working down the list, each strategy should be considered for its feasibility in the situation. The strategy with the most accuracy is recommended, if it is feasible in the situation.

Consider availability and convenience when selecting a strategy. Sometimes the availability of a particular source of data will drive the selection. In other situations, convenience may be an important factor in selecting a strategy.

When estimates are sought from individuals, use the source with the broadest perspective on the issue. Individuals asked to provide estimates must be knowledgeable about the processes and the issues relevant to the data.

Use multiple strategies when feasible. Sometimes it is helpful to have more than one strategy for obtaining a value for data. When multiple sources are available, more than one source should be utilized to serve as a comparison or to provide another perspective. When multiple sources are used, the data must be integrated using a convenient decision rule such as the lowest value. This is preferred because of the conservative nature of the lowest value.

Minimize the amount of time required to select and implement the appropriate strategy. As with other processes, it is important to keep the time invested as low as possible so that the total time and effort for the evaluation does not become excessive. Some strategies can be implemented with less time than others. Too much time at this step can dampen an otherwise enthusiastic attitude about the process.

ADDRESSING CREDIBILITY ISSUES

When converting data to monetary values, hard data can usually be converted in ways that are credible with the organization's stakeholders. However, when dealing with "output" as an improvement measure, it is always advisable to determine if the increased output is benefiting the organization through growth or increased sales. For example, if production improvements are made as a result of the training, and output increases substantially, you should determine if customer demand also exists for the output. If demand does not exist, then the product sits on a warehouse shelf taking up space, instead of providing increased revenue. When this happens, there may be a significant lag in benefits, and there is an immediate added cost of warehousing, which adds cost and thereby reduces the contribution of the increased output. Overlooking this may result in an overstated ROI and a credibility issue when stakeholders review evaluation results.

Soft data will sometimes be suspect with stakeholders. Therefore, when converting and presenting soft data it is important to follow strict principles such as the following:

- When organizational records are the source of data conversion, get several opinions on the credibility of the data in question.

- When people are asked to estimate values, be sure to make adjustments for confidence level.

- Always use the most conservative value when there is more than one possibility.

- After conversion, always round off to the nearest whole. For example, $120,035.00 becomes $120,000.00.

- When estimates are used in converting data, always communicate this fact to stakeholders.

MAKING ADJUSTMENTS TO THE DATA

Two potential adjustments should be considered before finalizing the monetary value of output. In some organizations where soft data are utilized and values are derived with imprecise methods, senior management is sometimes offered the opportunity to review and approve the data. Because of the subjective nature of this process, management may factor (reduce) the data so that the final results are more credible.

The other adjustment concerns the time value of money. Since an investment in a program is made at one time period and the return is realized in a later time period, a few organizations adjust the program output evaluation to reflect the time value of money, using discounted cash flow techniques. The amount of this adjustment, however, is usually small compared with the typical benefits realized from training and development programs.

CONVERTING A UNIT OF VALUE

The following steps should be taken to convert data to monetary values.

- Step 1. Determine the unit of improvement. *What measure is the program influencing? (E.g., increase in sales, decrease in scrap, time savings.)*
- Step 2. Determine the value of each unit (V). *What is the value of one unit of the measure? (E.g., one unit of sales, one unit of waste, one hour of time savings.)*
- Step 3. Determine the performance level change (ΔP). *How much did the measure change during the reporting period?*
- Step 4. Calculate the improvement value (V times ΔP).

The value of one unit of improvement (V) multiplied by the amount of change (ΔP) during the reporting period = the monetary value of the improvement. Figure 8.2 is a worksheet to assist in conversion to a monetary value.

Figure 8.2. (Downloadable form.)

Step 1.	Unit of improvement	_____
Step 2.	Value of each unit	_____
Step 3.	Performance level change	_____
Step 4.	Improvement value	_____

Using data from an example provided earlier in this chapter, a training team designed and implemented an intervention to reduce a turnover problem with chemical engineers. The intervention resulted in an annual reduction in turnover of 20 chemical engineers. Using Table 8.5, at an annual salary of $70,000, the value of retaining one engineer is conservatively $70,000. Figure 8.3 is an example of how the worksheet is used to develop a conversion.

Figure 8.3. Example of converting a unit of value.

Step 1.	Unit of improvement	Turnover
Step 2.	Value of each unit (cost)	$140,000 ($70,000 annual salary x 200%) using the lower percentage from Table 8.5
Step 3.	Performance level change	Twenty. Annual retention of 20 chemical engineers due to the intervention
Step 4.	Improvement value	$2,800,000 annually ($140,000 x annual retention improvement of 20 engineers)

ASKING THE RIGHT QUESTIONS ABOUT CONVERTING HARD AND SOFT DATA

The conversion of data is a critical step in the ROI process. In some instances, such as applying standard values that are routinely used by the organization to place a value on certain types of data, the process is consistent and credible. An example of this is a standard cost of quality, such as $300 per 100 rejects. However, when standard values are not available, such as with soft data and even some hard data, we must depart from standard values and apply numerous factors or criteria to assign values. Under these circumstances, the value chosen to use for conversion (the value of one unit of improvement—step two in Figure 8.3—can cause a significant swing in the return on investment. Extreme care should be exercised in selecting credible sources to convert data. Additionally, when a range of values is available, such as in Table 8.5, the most conservative value should be used unless stakeholders agree there is justification to use a different value. Figure 8.4 includes a worksheet with important questions to ask about converting hard data, and Figure 8.5 addresses soft data.

With the data converted to monetary values, the major steps in determining the benefits of a training program are complete. Collecting data, isolating the effects, and converting data to a monetary value are the major components. What remains now is capturing the costs of the program, calculating the ROI, and identifying intangible benefits, all of which are covered in upcoming chapters.

Figure 8.4. (Downloadable form.) Asking the right questions about converting hard data.

Worksheet: Questions to Ask About Hard Data

What is the value of one additional
unit of production or service?   _____

What is the value of a reduction of
one unit of quality measurement?
(rejects, waste, errors)   _____

What is the value of one unit of time
improvement?   _____

What are the direct cost savings?
(conversion not required)   _____

Figure 8.5. (Downloadable form.) Asking the right questions about converting soft data.

Worksheet: Questions to Ask About Soft Data

Are cost records available?   _____

Is there an internal expert who
can estimate the value?   _____

Is there an external expert who
can estimate the value?   _____

Are there any government, industry,
or research data available to estimate
the value?   _____

Are supervisors of program partici-
pants capable of estimating the value?   _____

Is senior management willing to
provide an estimate of the value?   _____

Does the training and development
staff have expertise to estimate the value?   _____

FURTHER READING

Phillips, Jack J. *Handbook of Training Evaluation and Measurement Methods*, 3rd Edition. Houston: Gulf Publishing, 1997.

Phillips, Jack J. 1994, 1997. *Measuring the Return on Investment*. Vol 1 and Vol 2. Alexandria, VA: American Society for Training and Development.

Phillips, Jack J. "Return On Investment in Training and Performance Improvement Programs." Houston, TX: 1997, Gulf Publishing.

Phillips, Jack J. "Was It The Training?" *Training & Development*, Vol. 50, No. 3, March 1996, pp. 28-32.

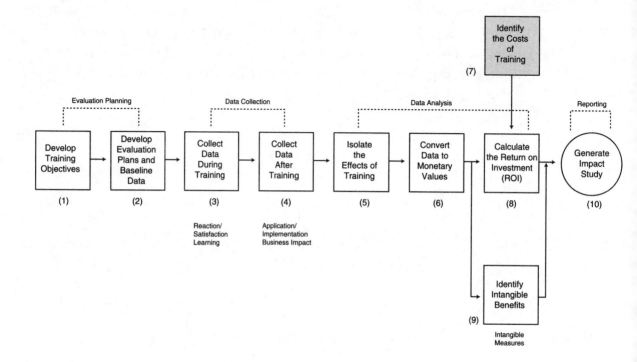

Identify
the Costs
of
Training

(7)

Evaluation Planning

Data Collection

Data Analysis

Reporting

Develop
Training
Objectives

(1)

Develop
Evaluation
Plans and
Baseline
Data

(2)

Collect
Data
During
Training

(3)

Reaction/
Satisfaction
Learning

Collect
Data
After
Training

(4)

Application/
Implementation
Business Impact

Isolate
the
Effects of
Training

(5)

Convert
Data to
Monetary
Values

(6)

Calculate
the Return on
Investment
(ROI)

(8)

Generate
Impact
Study

(10)

Identify
Intangible
Benefits

(9)

Intangible
Measures

9

Step 7. Identify the Costs of Training

IDENTIFYING THE COSTS OF TRAINING

This chapter describes the identification and tabulation of the costs of a training program, which specific costs should be considered, and some economical ways in which this can be done. It also presents useful worksheets. The good news is that costs generally are easier to identify than benefits. The issues to be addressed involve identifying the right costs and the fully loaded costs so that the resulting ROI calculation is a true representation.

THE IMPORTANCE OF COSTS IN DETERMINING ROI

One of the most important aspects of calculating the return on investment in training is identifying appropriate program costs. The relationship between the costs and benefits of a training program is direct. As illustrated in Figure 9.1, starting with example A, an increase in costs with the same level of benefits results in a decrease in the ROI, shown in example B. Likewise, as shown in example C, a decrease in costs with the same level of benefits will result in an increase in the ROI.

The cost figure must be accurate, reliable, and realistic. Although the total training budget is usually easily determined, it is more difficult to determine the actual cost of a specific program, including the indirect costs related to it.

Figure 9.1.

Example A	$\dfrac{\$300,000 - \$90,000}{\$90,000}$	$= 2.33 \times 100$	$= 233\%$
Example B Increase in costs with same level of benefits	$\dfrac{\$300,000 - \$120,000}{\$120,000}$	$= 1.50 \times 100$	$= 150\%$
Example C Decrease in costs with same level of benefits	$\dfrac{\$300,000 - \$50,000}{\$50,000}$	$= 5.00 \times 100$	$= 500\%$

Executives and other stakeholders recognize the potential swing in the ROI based on the cost factor. They are often concerned that an understatement of actual costs will inflate the ROI. As the examples indicate, a difference of 350% would be significant cause for concern when the reason for the difference might be an understatement of the costs. This is why it is important to fully load the costs and to be conservative in the process. When an ROI is presented, there should be no doubt that all appropriate costs are included in the calculation. This being the case, we can concentrate on the benefits side of the equation (numerator). We can address the credibility of our findings and our conversion of the benefits to monetary value without being distracted by the accuracy and credibility of the cost calculation.

DISCLOSING ALL COSTS

It is important to capture fully loaded costs. This goes beyond the direct costs of training and includes the time that participants are involved in training, including their salaries, benefits and other overhead. The internal competition to fund projects and other initiatives is driven by the effects of global competition. The need to improve products, market new products, stay competitive, and improve operational effectiveness is paramount in the minds of executives. So when executives think about the cost of training, they want to be assured that it will result in improvements and/or savings for the organization. Since funding is a scarce resource in most organizations, executives expect any claim for results to be objective and factual. An example will illustrate this.

A telecommunications company spent a significant amount of money to deliver an 80-hour middle-management leadership training program during business hours. When the results from an impact study were presented to a panel of senior executives, the ROI was 385% with the following costs reported:

- Instructor fees (external vendor)
- Instructor travel expenses
- Materials
- Meals and incidentals
- Participants travel expenses
- Research costs
- Valuation costs

One executive on the panel noted that the ROI seemed quite high. She inquired as to whether participant salaries and benefits were included. The evaluator responded that since middle managers were not directly involved in producing goods or services, there was no loss of output or other important organization results. She detailed how the managers would likely get most of their work done by working evenings or weekends and the organization would not see a loss. Another executive noted that there were no facilities costs included. The evaluator quickly responded that the program was conducted on company property and the facilities were there regardless of their use, therefore it would not be appropriate to include such costs.

A quick calculation by one of the executives revealed that by including salaries and benefits alone, the ROI would be less than half of the 385% reported. Others on the panel began to wonder what other costs might not be included and whether in fact the program had even achieved a positive ROI.

The manner in which the evaluator reported the costs in this example has several inherent flaws. First, the executive stakeholders should have had input into the nature of the costs to be included in the ROI calculation. This should have occurred during Step 2, the evaluation planning phase. Second, all costs should have been included. Taking any employee off the job is a cost. Likewise, the facilities cost should have been included. When organizations build, buy, or lease space, the need for meeting and training facilities adds to the floor space and furnishings required. Since we must pay for all square footage included in the facility, it is a cost that must be recovered over the life of the facility. Even when the exact costs are not known for employee salaries and benefits and space utilization, they can be estimated and included in the calculation. By failing to include all costs, we contribute to a debate over costs when we should be focusing our attention on the benefits of the training to the organization.

IDENTIFYING FULLY LOADED COSTS

With the conservative approach of fully loading costs, all costs that can be identified and linked to a particular program are included. The philosophy is simple: When there is doubt about a cost (the denominator of the ROI formula), include it, even if the cost guidelines for the organization do not require it. When the ROI is calculated and reported, the process should withstand even the closest scrutiny in terms of accuracy and credibility. The only way to meet this test is to ensure that all costs are included. Of course, from a realistic viewpoint, if a key stakeholder insists on not using certain costs, it is best to leave them out. Even then, the evaluator should encourage the stakeholder to include the cost.

PRESENTING COSTS WITHOUT PRESENTING BENEFITS

It is unwise to communicate the costs of training without presenting benefits as well. Many organizations have done this for years, and this mentality has contributed to a false expectation for some stakeholders that we can measure the impact of training by cost alone. Because costs can easily be identified, they are communicated in all types of ways, such as cost of the program, cost per employee, cost per contact hour, and cost compared to industry standard. While these methods may be helpful for efficiency comparisons, they have no relationship to results without the other part of the equation (benefits as shown in the numerator of the ROI formula). As stated earlier, when most executives review training costs, they are interested in the benefits received from the program. Some organizations have developed a policy of not communicating training costs unless the benefits can be presented along with the costs. Even if the benefit data is subjective and intangible, it should be included with the cost data. This helps to keep the frame of reference as it should be—what results were achieved from the targeted expenditure of funds.

It may be helpful to develop guidelines for the training function that communicate how training costs should be tracked and reported. Cost guidelines detail specifically what costs are included and how the data is captured, analyzed, and reported. Cost guidelines can be short or long, depending on the size and complexity of the organization. The simplest approach is best. When fully developed, the guidelines should be reviewed by the finance and accounting staff to ensure appropriate treatment. When an ROI is calculated and reported, costs are included in a summary form and the cost guidelines can be referenced in a footnote or attached as an appendix.

RECOMMENDED CATEGORIES FOR COSTS

The most important task is to define which specific costs are to be included. This requires the training staff to make decisions that may require approval by management. Table 9.1 shows the recommended categories for a fully loaded, conservative approach to estimating costs. Each category is described after the table.

Table 9.1. Cost categories.

COST ITEM	PRORATED	EXPENSED
Needs assessment	✔	
Design and development	✔	
Acquisition	✔	
Delivery		✔
■Salaries/benefits—trainers		✔
■Salaries/benefits—coordination		✔
■Program materials and fees		✔
■Travel/lodging/meals		✔
■Facilities		✔
■Participants' salaries/benefits		✔
■Contact time		✔
■Travel time		✔
■Preparation time		✔
Evaluation	✔	
Overhead/training and development	✔	

Prorated versus direct costs

Usually all costs related to a program are expensed to that program. However, needs assessment, program design and development, and program acquisition are usually prorated over several sessions of the same program. With a conservative approach, the shelf life of a program is very short. Some organizations will consider one year of operation for the program; others may consider two or three years. If there is some question about the specific time period to be used in the proration formula, the finance and accounting staff should be consulted.

A brief example illustrates the proration for development costs. In a large telecommunications company, a computer-based training program was developed at a cost of $98,000. It was anticipated that it would have a three-year life cycle before it would have to be updated. The revision costs at the end of the three years were estimated to be about one-half of the original development cost, or $49,000. The program would be conducted with 25 groups in a three-year period with an ROI calculation planned for one specific group. Since the program will have one-half of its residual value at the end of three years, one-half of the cost should be written off for this three-year period. Thus, the $49,000, representing half of the development costs would be spread over the 25 groups as a prorated development cost. So an ROI for one group would have a development cost of $1960, included in the cost profile (49,000 ÷ 25). Figure 9.2 provides a worksheet and illustration to help in prorating costs.

Figure 9.2. Worksheet—prorating costs.

Life Cycle of Program	*Number of Groups (offerings) During Life Cycle*	*Total Costs To Be Prorated*	*Total Costs Allocated to Each Group (offering)*
3 years	25	$100,000	$4,000
		$$\frac{\$100,000}{25} = \$4,000$$	
		$$\frac{\$}{\$} = \$$$	

Participant Salaries and Benefits. The salaries and employee benefits of participants should be included. Participant salary costs can be estimated using average or midpoint values for salaries in typical job classifications. When a program is targeted for an ROI calculation, participants can provide their salaries directly and in a confidential manner if this is considered necessary. Benefit costs usually are well known in an organization and are used in other costing formulas. The cost of employee benefits is usually expressed as a percentage of base salaries. In some organizations this value is as high as 50%–60%. In others, it may be as low as 25%–30%. The average in the United States is 38%.[1]

Needs assessment

One of the most often overlooked items is the cost of conducting a training-needs assessment. In some programs this cost is zero because the program is conducted without a needs assessment. However, as more organizations focus increased attention on needs assessment, this item will become more significant. All costs associated with the needs assessment should be identified to the fullest extent possible. These costs include the time of staff members conducting the assessment, direct fees and expenses for external consultants who conduct the needs assessment, and internal services and supplies used in the analysis. The total costs are usually prorated over the life of the program. Depending on the type and nature of the program, the shelf life should be kept to a very reasonable number in the one-to-two-year time frame. The exception would be very expensive programs that are not expected to change significantly for several years.

Design and development

One of the more significant items is the cost of designing and developing the training program. This includes internal staff time in both design and development; the purchase of supplies, videos, CD ROMs, and so on; e-learning development; administrative costs; consultant costs; and other costs directly related to the program. Design and development costs are usually prorated, perhaps using the same time frame as needs-assessment costs. One to two years is recommended unless the program is not expected to change for many years and the costs are very significant.

Acquisition

In lieu of program-development costs, many organizations have acquisition costs because they purchase training programs to use directly or in a modified format. Acquisition costs include the purchase price for the instructor materials, train-the-trainer sessions, licensing agreements, certification costs, and other costs associated with the right to deliver the program. Acquisition costs should be prorated like program-development costs. One to two years should be sufficient. If modification of the program is needed or some additional development is required, these costs should be included as development costs. Many programs have both acquisition costs and development costs.

Delivery costs

The largest segment of training costs usually is associated with delivery. Five major categories are included.

Salaries and benefits of trainers and coordinators. The salaries and benefits of the training staff involved with the program (e.g., trainers and coordinators) should be included. If a coordinator is involved in more than one program, time should be prorated to the specific program being evaluated. If external trainers are used, all their charges should be included. The important thing is to identify all time and expense costs for internal and external persons who are directly involved in producing the program. The benefits factor should be included each time direct labor costs are involved. This factor is a widely accepted value, usually generated by the finance and accounting staff and in the 30%–40% range.

Program Materials and Fees. Program materials such as notebooks, workbooks, textbooks, case studies, and job-aids should be included in the delivery costs, along with license fees, user fees, and royalty payments. Pens, paper, certificates, name tags, and calculators are also included in this category.

Travel, Lodging, and Meals. Direct travel expenses for participants, trainers, and coordinators should be included. Lodging and meals are included for participants during travel as well as during the program. Refreshments should also be included.

Facilities. The direct cost of the training facilities should be included. For external programs, this is the direct charge from the conference center, resort, or hotel. If the program is conducted in-house, the conference room represents a cost for the organization, and the cost should be estimated and included even if it is not the practice to include facilities costs in other reports.

Evaluation

Usually the total evaluation cost is included in the program costs to compute the fully loaded cost. ROI costs include the cost of developing the evaluation strategy, designing instruments, collecting data, analyzing data, and preparing and distributing reports. Cost categories include time, materials, and purchased feedback instruments (questionnaires and surveys). A case can be made to prorate the evaluation costs over several iterations of the program instead of charging the total amount as an expense. For example, if 25 sessions of a program are conducted in a three-year period and one group (offering) is selected for an ROI calculation, the ROI costs could logically be prorated over the 25 sessions. The results of the ROI analysis should reflect the potential success of the other programs and may result in changes that will influence the other programs as well.

Overhead

A final charge is the cost of overhead, the additional costs in the training function not directly related to a particular program. The overhead category represents any training department costs not considered previously. Typical items include the cost of clerical support, the departmental office expenses, salaries of training managers, and other fixed costs. Some organizations obtain an estimate for allocation by dividing the total overhead by the number of program participant days for the year. This becomes a standard value to use in calculations.

COST ACCUMULATION AND ESTIMATION

There are two basic ways to classify training costs. One is by a description of the expenditure such as labor, materials, supplies, travel, etc. These are expense account classifications. The other is by categories in the training process or function such as program development, delivery, and evaluation. These categories represent a type of activity where information is often needed to determine the relative costs throughout the training process. An effective system monitors costs by account categories according to the description of those accounts but also includes a method for accumulating costs by the training process/functional category. Many systems stop short of this second step. While the first grouping sufficiently gives the total program cost, it does not allow for a useful comparison with other programs or indicate areas where costs might be excessive by relative comparisons.

COST CLASSIFICATION MATRIX

Costs are accumulated under both of the above classifications. The two classifications are obviously related, and the relationship depends on the organization. For instance, the specific costs that comprise the analysis part of a program may vary from one organization to another.

An important part of the classification process is to define the kinds of costs in the account classification system that normally apply to the process/functional categories. Table 9.2 is a matrix that represents the categories for accumulating all training-related costs in the organization. Those costs, which normally are a part of a process/functional category, are checked in the matrix. Each member of the training staff should know how to charge expenses properly (e.g., equipment that is rented to use in the development and the delivery of a program). Costs will be allocated in proportion to the extent that an item is used for each category.

Table 9.2. Cost classification matrix.

| | EXPENSE ACCOUNT CLASSIFICATION | PROCESS/FUNCTIONAL CATEGORIES | | | |
		ANALYSIS	DEVELOPMENT	DELIVERY	EVALUATION
00	Salaries and benefits—training personnel	X	X	X	X
01	Salaries and benefits—other company personnel	X	X		
02	Salaries and benefits—participants		X	X	
03	Meals, travel, and incidental expenses—training personnel	X	X	X	X
04	Meals, travel, and accommodations—participants			X	
05	Office supplies and expenses	X	X		X
06	Program materials and supplies		X	X	
07	Printing and reproduction	X	X	X	X
08	Outside services	X	X	X	X
09	Equipment expense allocation	X	X	X	X
10	Equipment—rental		X	X	
11	Equipment—maintenance		X		
12	Registration fees	X			
13	Facilities expense allocation			X	
14	Facilities rental			X	
15	General overhead allocation	X	X	X	X
16	Other miscellaneous expenses	X	X	X	X

Reproduced from *Handbook of Training Evaluation and Measurement Methods,* Gulf Publishing, 1997, Jack J. Phillips, author.

COST ACCUMULATION

With expense account classifications clearly defined and the process/functional categories determined, it is easy to track costs on individual training programs. This is accomplished by using special account numbers and project numbers. An example illustrates the use of these numbers.

A project number can be a three-digit sequentially assigned number representing a specific training program. For example:

NEW PROFESSIONAL ASSOCIATES	
Orientation	112
New team leader training	215
Statistical quality control	418
Valuing diversity	791

Numbers are assigned to the process/functional breakdowns. Using the categories presented in Table 9.2, the following numbers are assigned:

Analysis	1
Development	2
Delivery	3
Evaluation	4

Using the two-digit numbers assigned to account classifications in Table 9.2, an accounting system is complete. For example, if workbooks are reproduced for the valuing diversity workshop, the appropriate charge number for that reproduction is 07-3-791. The first two digits denote the account classification, the next digit the process/functional category, and the last three digits the project number. This system enables rapid accumulation and monitoring of training costs. Total costs can be presented:

- By training program (e.g., valuing diversity workshop)
- By process/functional categories (e.g., delivery)
- By expense account classification (e.g., printing and reproduction)

COST ESTIMATION

The previous sections cover procedures for classifying and monitoring costs related to training programs. It is important to monitor and compare ongoing costs with the budget or with projected costs. However, another significant reason for tracking costs is to predict the cost of future programs. Usually this goal is accomplished through a formal cost estimation method unique to the organization.

Some organizations use cost-estimating worksheets to arrive at the total cost for a proposed program. Table 9.3 shows an example of a cost-estimating worksheet that calculates analysis, development, delivery, and evaluation costs. The worksheets contain a few formulas that make it easier to estimate the cost. In addition to these worksheets, current charge rates for services, supplies, and salaries are available. These data become outdated quickly and are usually prepared periodically as a supplement.

The most appropriate basis for predicting costs is to analyze the previous costs by tracking the actual costs incurred in all phases of a program, from analysis to evaluation. This way, it is possible to see how much is spent on programs and how much is being spent in the different categories. Until adequate cost data are available, it is necessary to use the detailed analysis in the worksheets for cost estimation.

When communicating the costs of a training program to stakeholders, it is extremely important to remember their frame of reference so that they do not view your costs as being out of line. For example, if you decide to communicate the cost of training programs on a cost-per-participant basis and you include participant salaries and benefits in the calculation, your stakeholders may view your training as expensive compared to the market. They may draw this conclusion because they are comparing the tuition of external programs with your per-person cost, which includes the additional cost of participant salaries and benefits. This results in an unfair comparison. It is important to explain this type of comparison so that others cannot take the costs out of context.

Table 9.3. Downloadable worksheet—cost guidelines.

	Itemized	Total
Program _____ *Date:* _____		
Analysis costs		
Salaries & employee benefits-training staff (number of people x average salary x employee benefits factor x number of hours on project)		_____
Meals, travel, and incidentals		_____
Office supplies and expenses		_____
Printing and reproduction		_____
Outside services		_____
Equipment expense		_____
Registration fees		_____
Other miscellaneous expenses		_____
Total analysis cost		═══════
Development costs		
Salaries & employee benefits-training staff (number of people x average)		_____
Salary x employee benefits factor x number of hours on project)		_____
Meals, travel, and incidental expenses		_____
Office supplies and expenses		_____
Program materials and supplies		_____
Film or videotape	_____	
Audio tapes	_____	
35-mm slides	_____	
CDs/diskettes	_____	
Overhead transparencies	_____	
Software	_____	
Artwork	_____	
Manuals and materials	_____	
Other	_____	
Printing and reproduction		_____
Outside services		_____
Equipment expenses		_____
Registration fees		_____
Other miscellaneous expenses		_____
Total development cost		═══════

Table 9.3. Worksheet—cost guidelines (*continued*).

Delivery costs   _____
Participant costs
Salaries & employee benefits-participants (number of
  participants x average salary x employee benefits factor
  x number of hrs or days of training time)   _____
Meals, travel, and accommodations(number of participants
  x avg. daily expenses x days of training)   _____
Participant replacement costs   _____
Lost production (explain basis)   _____
Program materials and supplies   _____
Instructor costs   _____
    Salaries and benefits   _____
    Meals, travel, and incidental expense   _____
    Outside services   _____
Facility costs (distance learning, traditional classroom, lab, other) _____
    Facilities rental   _____
    Facilities expense   _____
Equipment expense allocation
Other miscellaneous   _____

Total delivery costs   ═══════

Evaluation costs   _____
Salaries & employee benefits-training staff (number of people
  x average salary x employee benefits factor x number of
  hours on project)   _____
Meals, travel, and incidental expenses   _____
Participants costs (interviews, focus groups, questionnaires, etc.) _____
Office supplies and expenses   _____
Printing and reproduction   _____
Outside services   _____
Equipment expense   _____
Other miscellaneous expenses   _____

Total evaluation costs for program/project   ═══════

Share of general overhead allocation of training department   ═══════

Total program/project costs   ═══════

REFERENCES

[1]Annual Employee Benefits Report, *Nations Business*, January 1996, p. 28.

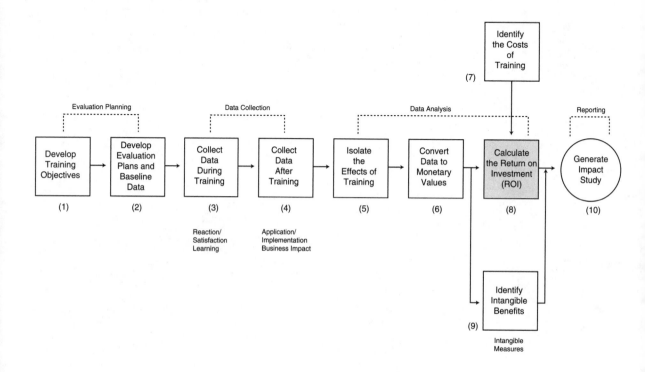

Identify
the Costs
of
Training

(7)

Evaluation Planning   Data Collection   Data Analysis   Reporting

Develop Training Objectives	Develop Evaluation Plans and Baseline Data	Collect Data During Training	Collect Data After Training	Isolate the Effects of Training	Convert Data to Monetary Values	Calculate the Return on Investment (ROI)	Generate Impact Study
(1)	(2)	(3)	(4)	(5)	(6)	(8)	(10)

Reaction/
Satisfaction
Learning

Application/
Implementation
Business Impact

Identify
Intangible
Benefits

(9)

Intangible
Measures

Step 8. Calculate the Return on Investment (Level 5)

CALCULATING THE BENEFIT-COST RATIO AND THE RETURN ON INVESTMENT

After data have been collected and analyzed for isolation and conversion, all that remains to calculate the benefit-cost ratio (BCR) and return on investment (ROI) is to apply the benefits to the numerator of the formula and compare it to the fully loaded costs of the program. The formula for benefit-cost ratio is stated as:

$$BCR = \frac{Benefits}{Costs}$$

For example, when the total benefits from a training program are $310,000.00 and the cost of the program is $92,000.00 using fully loaded costs, the BCR becomes:

$$BCR = \frac{\$310,000}{\$92,000} = \$3.36$$

This is expressed as 3.36: 1, or for every dollar spent, the benefit-cost ratio is $3.36.

The ROI formula is:

$$ROI = \frac{Net\ Benefits}{Costs} \times 100 = \underline{\qquad}\%$$

Applying the BCR example, when the total benefits from of the training program are $310,000.00 and the cost of the program is $92,000.00 using fully loaded costs, the ROI becomes:

$$\text{ROI} = \frac{\$310,000 - \$92,000}{\$92,000} = 2.36 \times 100 = 236\%$$

This yields an ROI of 236% after isolation and using fully loaded costs. Because executives and other stakeholders may not be accustomed to seeing training results expressed as a return on investment, they may question the credibility of the study. When presenting the results, care should be taken to provide a brief explanation of the methodology used to collect and analyze the data and calculate the ROI.

TEN GUIDING PRINCIPLES

The ten guiding principles are fundamental components of the ROI process. They are standards that help to ensure the implementation of a systematic and credible process to determine performance results. The ten principles are shown in Figure 10.1.

These principles should be discussed with management stakeholders during the evaluation planning process and when presenting the completed study.

THE POTENTIAL MAGNITUDE OF AN ROI FOR A TARGET POPULATION

The factors that influence a training program's bottom-line contribution can be categorized as shown in Figure 10.2.

By studying the factors in Figure 10.2, one can easily see why many training programs fail. Training is a process that includes many stakeholders, and some of them often are not aware of the roles they should play and the activities they should perform in order for a program to yield positive results for the organization.

A RATIONAL APPROACH TO ROI—KEEPING IT SIMPLE

The ROI process is not precise; neither are most other processes of this type. For example, estimates are used by engineers, scientists, production and quality control experts, and financial analysts.

Figure 10.1. Guiding principles.

1 When a higher-level evaluation is conducted, data must be collected at lower levels.
2. When an evaluation is planned for a higher level, the previous level of evaluation does not have to be comprehensive.
3. When collecting data, use only the most credible sources.
4. When analyzing data, select the most conservative alternative for calculations.
5. At least one method must be used to isolate the effects of the training.
6. If no improvement data are available for a population or from a specific source, it is assumed that little or no improvement has occurred.
7. Estimates of improvements should be adjusted for potential error.
8. Extreme data items and unsupported claims should not be used in ROI calculations.
9. Only the first year of benefits (annual) should be used in the ROI analysis of short-term programs.
10. The costs of a program should be fully loaded for ROI analysis.

Refinements are always possible. Our willingness to refine and be more precise is often tempered by cost, competitive resource allocation, and the time available to finish a project. While we always want to implement our process in credible ways, we have to balance resources and timeliness with our desire for more detail or more data.

Sometimes in order to satisfy someone's desire for more precise data, we can complicate the process unnecessarily. At some point, we must balance what is best for the project, for the organization, for the stakeholders, and for the credibility of the study. The checklist in Figure 10.3 can be helpful in keeping the project at a rational level and still satisfy the stakeholders and preserve credibility.

Figure 10.2. Factors contributing to ROI.

There must be a training need.	There must be a problem or opportunity where there is an existing performance gap or a new requirement is being introduced, and training is an appropriate solution.
A feasible training solution must be implemented at the right time, for the right people, at a reasonable cost.	The solution must be implemented based on the need. It should be implemented at a time that is compatible with the opportunity to perform or apply the new behavior. The program cost should be compatible with the cost or potential monetary benefit of the problem/opportunity. If the cost is out of line, this could result in a low or negative ROI.
The training solution must be supported and applied in the work setting.	Transfer to the performance setting must occur through application in order to influence individual and organizational performance. Plans should be implemented even before the training takes place to ensure a supportive application environment. This includes the appropriate involvement of numerous stakeholders to reinforce desired behavior. It also includes the existence of appropriate processes and systems.
Linkage must exist to one or more organizational measures.	If the behavior desired of training participants is to bring about organizational improvements, the training design must be linked to at least one organizational performance measure. This linkage should be established when the program is designed and communicated during participant selection and program implementation.

Figure 10.3. A rational approach.

- ✔ Keep the process simple.
- ✔ Use sampling for ROI calculations.
- ✔ Always account for the influence of other factors.
- ✔ Involve management in the process.
- ✔ Educate the management team.
- ✔ Communicate results carefully along with the methodology.
- ✔ Give credit to participants and managers.
- ✔ Plan for ROI calculations.

ROI COMPARISONS

While it is tempting to compare ROI calculations for similar programs, the practice should be discouraged. As illustrated in Figure 10.2, many factors must be in place to optimize the contribution of training. If comparisons are to be made, they should be made regarding how best practices can improve linkage to each phase of the training process as well as each phase of the evaluation process. The ultimate ROI will be positively influenced if we continue to focus on training process improvement. When a specific ROI study reveals an undesirable outcome, we should immediately use that discovery to influence training process improvement instead of being concerned with numeric comparisons. Evaluation is not a numbers game. It is a process of discovery, replication of success, and correction of shortcomings.

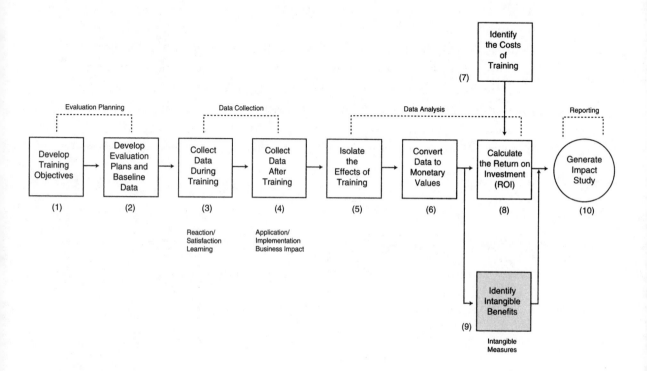

Identify
the Costs
of
Training

(7)

Develop
Training
Objectives

(1)

Develop
Evaluation
Plans and
Baseline
Data

(2)

Collect
Data
During
Training

(3)

Reaction/
Satisfaction
Learning

Collect
Data
After
Training

(4)

Application/
Implementation
Business Impact

Isolate
the
Effects of
Training

(5)

Convert
Data to
Monetary
Values

(6)

Calculate
the Return on
Investment
(ROI)

(8)

Generate
Impact
Study

(10)

Identify
Intangible
Benefits

(9)

Intangible
Measures

11

Step 9. Identify Intangible Benefits

WHY INTANGIBLE BENEFITS ARE IMPORTANT

Most successful training programs result in some intangible benefits. Intangible benefits are those positive results that either cannot be converted to monetary values or would involve too much time or expense in the conversion to be worth the effort. The range of intangible outcomes is practically limitless.

Intangible benefits should be measured and reported. They can be used as additional evidence of a training program's success and can be presented as supportive qualitative data. Intangibles may not carry the weight of measures that are expressed in dollars and cents, but they are still a very important part of the overall evaluation, and many executives are interested in these measures.

It is important to note that, even though any of the intangible benefits may not be converted in one evaluation study, they may be converted in another study or in another organization. In some situations, intangible effects on teamwork, job satisfaction, communication, and customer satisfaction are as important as monetary measures.

IDENTIFYING COMMON INTANGIBLE VARIABLES

The intangible variables listed in Figure 11.1 are the more common ones.

Figure 11.1. Common intangible benefits.

■ Increased job satisfaction	■ Decreased customer dissatisfaction
■ Increased organizational commitment	■ Enhanced community image
■ Improved work climate	■ Enhanced investor image
■ Fewer employee complaints	■ Fewer customer complaints
■ Fewer employee grievances	■ Faster customer response time
■ Reduction of employee stress	■ Increased customer loyalty
■ Increased employee tenure	■ Improved teamwork
■ Reduced employee lateness	■ Increased cooperation
■ Reduced absenteeism	■ Reduction in conflict
■ Reduced employee turnover	■ Improved decisiveness
■ Increased innovation	■ Improved communication
■ Increased customer satisfaction	

There are typically two types of intangibles: behavioral application/implementation (Level 3) and business impact (Level 4). Behavioral intangibles are Level-3 results that are influenced by a training program. Examples are improved teamwork, increased organizational commitment, increased cooperation, and improved communication. If the evaluation process can link these results to Level-4 measures, they become Level-4 intangibles. If we can convert the Level-4 measures to monetary values, we have succeeded in making these measures more tangible. If the process is credible, this makes the measures more objective and, if we choose, we can calculate the ROI.

Examples of Level-4 intangibles are customer satisfaction, employee satisfaction, fewer employee complaints or grievances, reduced employee absenteeism, and reduced employee turnover. If we can place a value on any of these with credibility, we can move it from an intangible to tangible category. The factors that influence success in placing value on intangible measures range from the difficulty in arriving at a credible value to a lack of need from stakeholders. The following example may help to clarify this.

A training program in customer service skills has an objective of improving three measures in Customer Service Center (CSC) operations:

1. Resolve customer complaints/issues at the associate level.
2. Learn how to use the systematic customer-engagement process (SCEP) and apply it in all customer-contact situations.
3. Learn and apply the new team-support process (TSP) to ensure that customer complaints are handled and resolved properly when a customer call must be passed from one associate to another.

Following the training, our evaluation reveals that the CSC associates are applying the systematic customer-engagement process and the team-support process on a regular basis (90% of the time). This has greatly improved communication and teamwork. As a result, improvements attributed to the training include:

- A 50% reduction in customer calls that are referred to the CSC supervisor for resolution
- A 70% improvement in dropped calls (initial calls and internally transferred calls that result in the customer call being terminated without resolving the complaint)

The fact that associates in the CSC are applying these proven processes on a regular basis (90% of the time) is a level-3 intangible. Other level-3 intangible benefits are increased teamwork and communication. The 50% reduction in customer calls that reach the CSC supervisor and the 70% improvement in dropped calls are both level-4 intangible measures of improvement. Unless the stakeholders are interested in an ROI, we could report these level-4 improvements as intangible results. We do this because we are unable to determine if the benefits exceed the costs of the program and, if so, by how much.

We could have spent too much or we may have been able to spend less and still achieve an acceptable result. In order to develop the ROI calculation, we would need to determine a credible way to place a monetary value on the two level-4 measures. Although this may be possible, the stakeholders may be satisfied with the intangible results and may not be interested in an ROI calculation. They may be satisfied that the expenditure was wise and the desired result achieved. This may be particularly true if the expenditure (cost of the training) is a relatively small amount of money compared to the perceived benefit.

SOURCES FOR INTANGIBLE BENEFITS

During needs assessments. While conducting needs assessments, there are numerous opportunities to address intangibles. Many needs often start at level 3, and as we begin to question how behavior influences outcomes, we can identify level-4 intangible needs as well. As we identify needs, we should attempt to place a monetary value on the business problem/opportunity. For example, if we identify a level-4 need to reduce excessive employee absenteeism, we have identified a level-4 intangible. If we can place a value on absenteeism, we can calculate what it is costing the organization (cost of one absenteeism x number of excessive absences = cost to the organization). We now have a tangible value in management terms.

During early planning. During the planning phase of the evaluation process, when linking objectives to business measures or addressing baseline measures, intangibles are often uncovered. The objectives of the training may also reveal intangibles.

Clients. Clients of the training function often can identify intangibles. A client who seeks training assistance often identifies a need in level-3 language, such as: "My associates are not working together as a team," or "Our lack of communication is hurting our ability to serve our customers." Likewise, clients can identify when these intangible measures have improved.

Participants. Participants may be one of our best sources in identifying intangibles. Since it is their performance we are trying to influence, they can tell us how things change for them in the organization as they implement new skills.

Managers. Managers often have a broader view of the work setting and can see the overall behavioral changes and how they impact important intangible measures.

Customers. Customers often identify intangibles when they discuss the performance of an associate—comments like "She is so friendly and helpful," or "I feel like my account has been handled with care" or "Your office displays such a high degree of professionalism." Often, we can link such behavior to improved customer satisfaction.

During data collection. We often uncover intangible data during data collection from stakeholders. We may not always plan to collect intangible data, and we may not anticipate it in the initial planning. However, intangible data may surface on a questionnaire, during an interview, or during a focus group activity. When questions are asked about improvements influenced by the training, participants usually provide several intangible measures for which there are no plans to assign a value.

During conversion attempts. Intangibles are often identified when attempts at conversion are aborted. When level-4 data cannot be converted to a monetary value (or we choose not to convert), we are inherently left with intangible outcomes. When this occurs, we should explain the reason we are unable to convert.

A FINAL WORD ON INTANGIBLE BENEFITS

It is important to identify trends when capturing intangible benefits. It is tempting to use sparse comments or observations by clients, customers, participants, or supervisors of participants to draw conclusions about the intangible benefits resulting from a program. One positive comment about one associate from a customer or supervisor may be insufficient to claim an overall level-3 intangible benefit from the program. An objective evaluator will look further and deeper for intangible benefits.

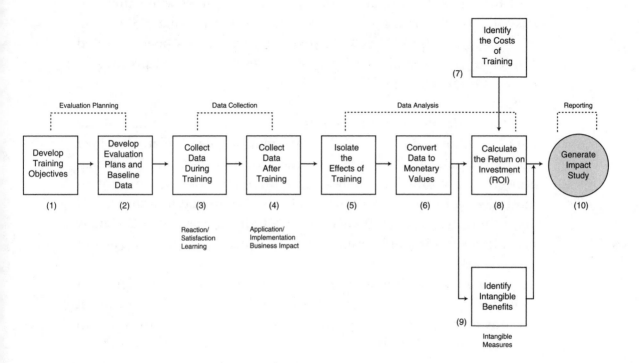

12

Step 10. Generate an Impact Study

THE NEED TO IDENTIFY ORGANIZATIONAL IMPACT

With increased competition for organizational resources, there is increased pressure to determine the contribution that training makes to an organization. Executives make decisions to allocate funds for training just as they make decisions to fund other initiatives and other departments. They have a keen interest in the payback resulting from the funding.

This point is illustrated by a situation that recently developed in a large electric utility holding company. The holding company had been funding a corporate college for about 10 years. As the threat of deregulation surfaced, the industry began to restructure in reaction to competitive pressures and regulatory requirements. The company created a new chief operating officer (COO) position and placed a well-respected line officer in the position. The new COO had a long-standing reputation for scrutinizing costs and focusing on the bottom line. He began questioning the costs associated with the college, and the dean of the college was unable to show results for the dollar expenditures. Participant smile sheets and staying within the college's allocated budget did not spell "accountability" to the COO concerned with competitive product pricing in a rapidly changing industry.

The dean had several opportunities to track costs and results, compare these results to the funding, and present business impact reports and other results to the COO. Instead, the dean listened to his design staff and a Ph.D. he had recently hired to head the evaluation function. They convinced the dean that smile sheets and other reaction data pointed to the impact the college was making. The COO had no information that demonstrated a contribution compatible with the funding of the college. In short order, a few of the college staff were reassigned to functions that management could support, while the dean and the bulk of the college staff accepted outplacement packages. This story is waiting to be repeated throughout corporations and government agencies. One way to change the ending is to provide the right kind of data to stakeholders and to install accountability processes and measures in the training function.

MONITORING PROGRESS

A final part of the ROI process is to monitor the overall progress made and communicate the results. Although it is an often-overlooked part of the process, an effective communication plan can help keep the process on target and let others know what the ROI process is accomplishing for the organization.

As training programs are implemented and the ROI evaluation is underway, status meetings should be conducted to report progress and discuss critical issues with appropriate team members. For example, if a supervisory training program for operations is selected as one of the ROI projects, all of the key staff involved in the program (design, development, and delivery) should meet regularly to discuss the status of the evaluation project. This keeps the project team focused on the critical issues, generates the best ideas to tackle particular problems, and builds a knowledge base to implement evaluation in future programs. This group may be facilitated by an internal ROI leader or an external consultant who is an expert in the ROI process.

These meetings serve three major purposes: reporting progress, learning, and planning. The meeting usually begins with a status report on the ROI project, describing what has been accomplished since the previous meeting. Next, the specific problems encountered are discussed. During the discussions, new issues are interjected in terms of possible tactics, techniques, or tools. Also, the entire group discusses how to remove barriers to success and focuses on suggestions and recommendations for next steps, including developing specific plans. Finally, the next steps are determined.

FOCUSING ON CONTRIBUTION, NOT JUSTIFICATION

As measurement takes place to demonstrate results, the focus should be on the contribution made by the entire process. A measurement objective that seeks to justify the funding for training or the existence of training is inappropriate and will usually fail. The results should be clearly communicated. The training department or corporate university should not take singular credit for the improvement.

Each phase of the process should be addressed in the communication of results. In addition, as numerous studies are completed, the strengths and weaknesses of the various phases of the process should be addressed. Thorough analysis of results can reveal how the strengths and weaknesses affect business impact (level 4) and ROI (level 5). This information then can be used to focus resources and funding on the various phases of the process that require reinforcement.

COMMUNICATING RESULTS

There are many potential target audiences for communication of ROI evaluation data on the results of training. However, four particular audiences should always receive ROI data. The senior management team (however it may be defined) should always receive information about the impact of training because of its interest in the process and its influence in allocating additional resources for training. The supervisors of program participants need to have the ROI information so they will continue to support programs and reinforce specific behaviors taught in the program. The participants in the program who actually achieved the results should receive a summary of the ROI information so they understand what was accomplished by the entire group. This also reinforces their commitment to make the process work. The training staff must understand the ROI process and should receive the ROI information from each program as part of the continuing educational and improvement process.

ADDRESSING THE NEEDS OF TARGET AUDIENCES

A single report used for all audiences generally is not appropriate. Although there may be substantial data to report, it is important for each communication to be tailored directly to the needs of the particular audience and the objectives of the training department with that audience. Planning and effort are necessary to make sure that each audience receives all of the information it needs and is interested in (the type of information, the focus of the information, and the scope of the information), in the proper format (the size of the report and the media used), at the proper time. These factors may vary significantly from one stakeholder group to another.

Executives usually require brief summaries of the background, findings, and recommendations. Participants may not be as interested in the ROI, but have interest in the parts of the study that address application in the work setting. The training staff usually has an interest in the entire report, including the details.

DEVELOPING THE EVALUATION REPORT

One of the most important documents for presenting data, results, and issues from an ROI impact study is an evaluation report. The typical report provides background information, explains the processes used, and presents the results. Recognizing the importance of on-the-job behavior change, level-3 results are presented first.

Business impact results are presented next, which include the actual ROI calculation. Finally, other issues are covered along with the intangible benefits. While this report is an effective and professional way to present ROI data, there are several caveats. Since this document is reporting the success of a training program involving a group of employees outside of the training function, credit for the successful implementation results must go to the participants and their immediate supervisors, whose performance has generated the success. Also, it is important to avoid boasting about the results. Although the ROI process may be accurate and credible, it still may have some subjective issues. Huge claims of success can quickly turn off an audience and interfere with the delivery of the desired message.

A final caution concerns the structure of the report. The methodology should be clearly explained along with the assumptions made in the analysis. The audience should readily see how the values were developed and how the specific steps were followed to make the process more conservative, credible, and accurate. Detailed statistical analyses should be relegated to the appendix.

In some situations, particularly when training initiatives require extensive funding, the amount of detail in the evaluation report is more crucial. A complete and comprehensive impact report may be necessary. This report can then be used as the basis of information to prepare more focused reports for specific audiences.

A comprehensive report may contain the key areas outlined in Figure 12.1.

DEVELOPING THE IMPACT STUDY

Management/executive summary. The management summary is a brief overview of the entire report, explaining the basis for the evaluation and the significant conclusions and recommendations. It is designed for individuals who are too busy to read a detailed report. It is usually written last but appears first in the report for easy access.

Figure 12.1. Impact study—Outline of report format.

General Information
- Background
- Objectives of study

Methodology for Impact Study
- Levels of Evaluation
- ROI Process
- Collecting Data
- Isolating the Effects of Training
- Converting Data to Monetary Values

Data Analysis Issues

Costs

Results: General Information
- Response Profile
- Success with Objectives

Results: Reaction/Satisfaction, Planned Actions
- Data Sources
  Data Summary
- Key Issues

Results: Learning
- Data Sources
- Data Summary
- Key Issues

Results: Application and Implementation
- Data Sources
- Data Summary
- Key Issues

Results: Business Impact
- General Comments
- Linkage with Business Measures
- Key Issues

Results: ROI Calculation and Its Meaning

Results: Intangible Benefits

Barriers and Enablers
- Barriers
- Enablers

Conclusions and Recommendations
- Conclusions
- Recommendations

Exhibits

Background information. The background information provides a general description of the training program. If applicable, the needs assessment that led to the implementation of the training is summarized. The solution is fully described, including the events that led to the training program and the overall objectives. A full description (synopsis) of the training is provided, including features of the training design. The required level of detail and extent of this information depends on the audience.

Objectives of the study. The report details the objectives of the evaluation study so that the audience clearly understands its purpose and desired accomplishments. In addition, the issues or objectives from which the different types or levels of data will be collected are detailed here.

Evaluation methodology and strategy. The evaluation strategy outlines all the components of the evaluation process. Several components of the results-based model and the ROI process presented in this book are discussed in this section of the report. The evaluation design and methodology are explained. The instruments used in data collection are also described and presented as exhibits. Any unusual issues in the evaluation design are discussed. Finally, other useful information related to the design, timing, and execution of the evaluation is included.

Data collection and analysis. This section explains the specific methods used to collect data (how and when). The data collected are usually presented in the report in summary form. Next, the methods used to analyze data are presented. Finally, the resulting interpretations are presented.

Program costs. Program costs are presented and summarized by category. For example, needs analysis, program development, program implementation, and evaluation are some recommended categories for presentation of costs. The assumptions made in developing and classifying costs are discussed in this section.

Reaction and satisfaction. This section details the data collected from key stakeholders, particularly the participants involved in the training process, to measure the reaction to the training and the level of satisfaction with various issues and parts of the process. Other input from the client group is also included to show the level of satisfaction. If planned applications are part of the data, they are presented here.

Learning. This section shows a brief summary of the formal and informal methods for measuring learning. It explains how participants have learned new behaviors, procedures, processes, tasks, and skills from the training.

Application and implementation. This section shows how the training was actually applied or implemented on the job and the success of the application of the new knowledge and skills. Implementation issues are addressed, including any major success and/or lack of success.

Business impact. This section shows the actual business impact measures and how they relate to the business needs that drove the training initiative. This section shows the extent to which performance has changed as a result of the implementation of the training.

Return on investment. This section shows the ROI calculation along with the benefits-cost ratio. It compares the value to what was expected and provides an interpretation of the actual calculation. It also briefly reinforces the key strengths of the methodology that was used to arrive at the calculation. For example, it mentions that the analysis used conservative approaches, that extreme data and unsupported data were not used in the calculation, and that the costs were fully loaded.

Intangible measures. This section shows the various intangible measures directly linked to the training. Intangibles are those measures not converted to monetary values and not included in the ROI calculation.

Barriers and enablers. The various factors or influences that had a positive effect on the implementation of the training (enablers) are identified. Any problems and obstacles that adversely affected the implementation of the training (barriers) also are detailed. This section of the report can provide tremendous insight into what can enhance or hinder training and performance-improvement initiatives in the future.

Conclusions and recommendations. This section presents conclusions based on all the results. If appropriate, a brief explanation is presented of how each conclusion was reached. If appropriate, a list of recommended changes in the program also is provided, with a brief explanation of each recommendation. It is important that the conclusions and recommendations be consistent with one another and with the findings described in the previous sections of the report.

These components make up the major parts of a complete evaluation report. It is important to keep the report to a minimum amount of information that will satisfy the audience. In addition, the appropriate use of media should be considered. For example, some possible media include the employee newsletter, a special meeting with computerized imaging, and even Web conferencing when multiple geographic locations and timeliness are issues.

PRESENTING A BALANCE OF FINANCIAL AND NONFINANCIAL DATA

While executives are almost always interested in financial data such as ROI, they also want to know what drives the numbers. In the case of a successful training program, the numbers are driven by people who learn a new skill or exhibit new attitude, then apply the behavior that drives desired business outcomes. In many situations, we are unable to convert data to a monetary value and therefore we do not have financial data to report.

The typical study includes both financial data and data that is not converted to a monetary value (nonfinancial or intangible data). As discussed in Chapter 11, it is important to report both types of data. There have been many situations where a training program yields a disappointing ROI, yet the intangible data showed that the training made a significant contribution. We should not lose sight of the definition of ROI. It simply answers the question, "Did we get back more (in financial terms) than we paid for the program, or did the funding exceed the benefits?"

This question is asked of most business initiatives. However, in many situations, the rigor of an ROI study is not pursued. Why then, do we choose to report the results of training in ROI terms? There are several reasons. First, many of the executives who fund training are not convinced that training yields results, thus they require ROI information. Second, unless we can make training available during off-company hours (using any type of delivery channel), the work time in which people are engaged in training (through whatever media) is a cost to the organization, and the participants are not engaging in the work activities that support the organizations' products and services. Third, line managers may not be convinced that training provides benefits to the work setting. We need to develop and communicate level-3 objectives for training programs in terms that line managers can identify with and understand, and we need to show them level-3 results. In addition, we need to partner with managers and others to create structures and attitudes that allow transfer of learning to succeed in the work setting.

Finally, we need to address the numerous assumptions about the training process and training results that float around organizations. For example, much training does not train to specific competencies. This may be because of insufficient funding or it may be because of insufficient hours of participant availability. But line managers and other stakeholders may have the expectation of "full competency" when people complete a training program. By allowing this standard to be formulated in the minds of our stakeholders without our input, we allow a false expectation to exist. This often results in the conclusion that we are not meeting needs and therefore not achieving results. Stakeholders may not be aware that the funding and policies to support the standard may not be in place. So we spend a lot of time trying to convince stakeholders that we are achieving results.

As training professionals, we need to focus our efforts on establishing expectations first and then build everything else around these expectations. The five levels of evaluation provide an excellent framework for creating expectations and evaluating and communicating results. Figure 12.2 includes examples of some questions we should address. Some of the questions may be asked of all stakeholders. The training staff should commit quality time to identifying their stakeholders and customizing these types of questions for the stakeholders in their organization. This will not only provide direction but will also make evaluation responsibilities easier to address.

Figure 12.2. (Downloadable form.) Stakeholder expectations.

Example: Questions To Ask To Determine Stakeholder Expectations			
LEVEL	**EXECUTIVE**	**LINE MANAGER**	**PARTICIPANT**
Level 1, Reaction/ Satisfaction Planned Action	What do you view as participant responsibilities when they attend training programs?	What are your preferences (time, location, etc) as your associates attend our training programs?	What is important to you as you participate in our programs? What do you expect of the experience?
Level 2, Learning	What delivery channels will you support to achieve learning in the organization? What funding is available for training? What funding is available for experimentation with the various delivery channels?	What level of learning do you expect? How much time will you allocate for you and your associates to allow this to happen? How can you be involved?	What do you need to learn? How will this benefit you? How would you like to learn? What learning methods work best for you?
Level 3, Application/ Implementation	What should people be doing to contribute to achieving strategic objectives? How can you demonstrate support for learning transfer? How can we assist in that? What funding is available to influence transfer?	What should your associates be able to do after attending our training program? How can you become involved before and after the training to make this happen?	What do you need to be able to do? How can we best help you learn to do that? What enablers need to be in place to help you do these things?
Level 4, Business Impact	What problems or opportunities exist that we can influence with training? What organizational measures need to be influenced by training programs? What evidence do you need that will demonstrate that results have been achieved?	If your people apply what they learn, how will it benefit the organization? What are the other factors that can also influence the results you want?	When you apply the new skills/behavior identified, how will the organization benefit? What measures will improve?

Figure 12.2. Stakeholder expectations (*continued*).

Level 5, Return on Investment	To what extent do you expect the benefits of the training to exceed the fully loaded cost of the training (what ROI is acceptable)?	How will the program benefit your operation and provide a return on the participants' time and any other investment or lost opportunity?	How will the program personally benefit you and provide a return on your time and any other investment or lost opportunity?

Copyright McGraw-Hill 2002. To customize this handout for your audience, download it from (www.books.mcgraw-hill.com/training/download). The document can then be opened, edited, and printed using Microsoft Word or other word-processing software.

Getting the answers to these sample questions (and others) will allow you to do several things:

- Conduct the proper research
- Establish the appropriate expectations
- Set objectives to meet expectations
- Evaluate programs to determine success and contribution
- Report results data
- Be accountable

In short, discovering the expectations of all stakeholders can assist tremendously in meeting those expectations.

What to do if you get embarrassing or disappointing measurement data

Sometimes training programs or other interventions simply do not produce results. Reasons for failure can range from the fact that the learning needs of participants were not addressed to the conclusion that the work environment was not supportive. Just as it is important to identify why programs are successful so that we may replicate the process, it is important to identify what went wrong so we can improve the training process and follow up. A failed intervention should become the basis of a needs assessment for continuous improvement.

When reporting on a program that does not meet expectations, it is important to view the results from several perspectives.

- Was the training need properly identified?
- Were the training objectives aligned with the need?
- Were the right people involved in the training process, including participants and other stakeholders?
- Was the timing of the training appropriate?
- Was the proper content delivered?
- Was there sufficient skill practice?
- Was the training application/implementation supported in the work environment?
- Was the training aligned with appropriate business measures?

It also is important to examine both tangible and intangible data. Occasionally, even though the tangible data may show that expectations are not met, an examination of the intangible evidence may show many benefits to the organization. For example, a major engineering company set an expectation of an ROI of 50% for a key leadership program. The program fell short of expectations when the monetary benefits of the results, compared to the cost of the program, showed an ROI of only 10%. However, an examination of the intangible data revealed that several benefits could not be converted to monetary values. Employees reported they felt like they had more direction and significantly better coaching since their managers had participated in the training. The managers reported that they were able to plan the work week with greater efficiency and they had more free time to concentrate on customer and product issues. While both results are intangible, senior management decided that the intangibles far outweighed the low ROI and concluded that the program was successful.

When a program does not meet expectations, the reasons for the deficiency should be identified and disclosed to management and other stakeholders. Improvement steps should be identified and recommended. It is important to communicate the results carefully, as miscommunication can cause major problems. Because the results of a training program can be closely linked to the political agendas in an organization, communication can upset some individuals while pleasing others. If certain individuals do not receive the information or it is delivered inconsistently from one group to another, problems can quickly surface. Not only is it an understanding issue, it is also a fairness, quality, and political correctness issue to make sure communication is properly constructed and effectively delivered to all key individuals who need the information.

GIVING OTHERS DUE CREDIT

Training is an enabling process: it consists of identifying needs and designing ways to provide participants with the necessary tools to improve their performance in desired ways. When participants seek to apply what they learn in the training, the organizational setting must encourage and reinforce the application of desired knowledge and skills. The result should be visible outcomes that benefit the organization. A positive ROI is the result of a chain of events involving many stakeholders, and it is important to give credit for a successful outcome to all of them. There are many people involved in the process, including:

- Those who contribute to the needs assessment and properly identify needs and opportunities
- Those who champion, approve and fund the training effort
- Those who design the training to meet the desired objectives and create or obtain the appropriate training materials
- Those who present the training and implications for application in a way that furthers the desired objectives
- The training participants

Figure 12.3.

		ROI Contribution ⟶			
STAKEHOLDERS	PARTICIPATES IN IDENTIFYING NEEDS AND PERFORMANCE GAPS	PROVIDES FOR TRAINING DELIVERY AND FACILITATES LEARNING	PARTNERS TO PLAN AND SUPPORT LEARNING TRANSFER	IMPLEMENTS BEHAVIOR FROM LEARNING THAT DIRECTLY LEADS TO RESULTS	ULTIMATE SOURCE OF FUNDING
Training Staff	yes	yes	yes	no	no
Client Supervisors and Managers	yes	sometimes	yes	no, but must support the behavior	sometimes
Participants	yes	Yes, facilitates learning	yes	yes	no, unless it is paid out of participant's pocket
Co-workers of participants	sometimes	sometimes	sometimes	no, but must support the behavior	no
Senior Managers	sometimes	sometimes	sometimes	no, but must support the behavior	yes
Frequent designated players such as experts, or thought leaders	sometimes	sometimes	sometimes	no	no

- The participants' managers and peers who continue the ongoing productivity of the organization while the participants are completing the training
- Those who help to design the evaluation criteria, collect pre- and post-training data, assess the data, isolate the effects of the training, convert the data to monetary values, and calculate the return on investment
- Those who support the participants' attempts to apply/implement what they have learned in the work setting and those who create structures to reinforce this performance improvement

Figure 12.3 helps to illustrate some of these roles.

When reporting a successful outcome, it is important to recognize those who participated in less visible ways—such as those who collect and analyze data or convert data to monetary values and the managers who support data collection and other aspects of the training and the evaluation process.

In the end, it is the participants who achieve the results. The participants must implement the learned behaviors or there will be no result. The process reaches closure when appropriate partnering occurs to ensure results in the organizational settings. It is inappropriate for the training staff to take credit by claiming that the training program caused the results. The truth is, the training process causes the results. The process (not the program or the training department) should get the credit, and the focal point should be on the participants and their managers.

13

Fast and Easy Approaches to Measuring Training Outcomes

As the ROI process is planned, determining the methods to be used for the various components of the process is a crucial decision that affects the resources required to conduct the evaluation. As long as you are not compromising the process, you should always look for efficient ways to achieve the evaluation objectives. This chapter addresses cost-saving and quick approaches that it may be possible to utilize when planning and conducting a training evaluation.

COST-SAVING APPROACHES TO ROI

Plan for evaluation early in the process. Build evaluation in at the needs-assessment and program-design steps. Even if the evaluation is not pursued, the cost to build it in early is minimal, and it will be ready if someone decides that the program should be evaluated.

Partner with designers and vendors to build-in the process. Ask designers to work with you to build-in evaluation strategies. Require vendors to become involved and seek their funding for the process

Build evaluation into the training process. Evaluation on a sampling basis should become a routine expectation by trainers, participants and others involved in the process. If unions or legislative requirements inhibit evaluation, build in anonymity and seek cooperation of third parties.

Share the responsibilities for evaluation. Delineate responsibilities to trainers and coordinators for level-1 and level-2 evaluation. Seek additional help at the other levels and make it a learning experience.

Require participants to conduct major steps. Participants are a rich source of data, and they can be useful in collecting data when properly instructed. This can be accomplished with questionnaires, action planning, performance contracts, follow-up assignments, and other methods.

Use sampling to select the most appropriate programs for ROI analysis. Create meaningful strategies on why and how you will proceed with evaluation. Then put a plan in place to sample what is happening throughout the training process. Remember the guiding principle that, when data is collected at higher levels, the data at lower levels does not need to be comprehensive.

Use estimates in the collection and analysis of data. Estimates are used throughout the organization by operations, accounting, finance, marketing, engineering, and production/manufacturing. While estimates are not the preferred method for collection and analysis, they can be very credible when collected from the right sources and when adjusted for potential error.

Develop internal capability to implement the ROI process. Utilize reading materials and job-aids, and participate in skill-building workshops and networking to sharpen skills and seek out best practices.

Utilize Web-based instruments, electronic data scanning, software, and templates to reduce time. Web-based techniques can be utilized to save considerable time in questionnaire administration and telephone interviews. Scanning and data analysis software can be utilized to significantly reduce the time to administer studies and analyze data. Templates can be used to plan, design questionnaires, and implement studies.

Streamline the reporting process. Use templates and present only relevant data to each audience. Some stakeholders, such as senior managers, may be interested in levels-3, -4, and -5 data but may have limited interest in levels-1 and -2 data. Make your reports fit your audience and don't overdo it. More is not always better. Consider developing a template that describes how the ROI process works and place it in the appendix of the report.

THREE QUICK WAYS TO DISCOVER WHAT HAPPENED, WHY, AND WHAT THE RESULT IS WORTH

As discussed in Chapter 2, the ROI process includes three principal follow-up components: collecting data, isolating the effects, and converting data. The options for capturing the costs of the program are also a factor in resource utilization. However, this component usually takes the least amount of time. The resource requirements (time and cost) to conduct an evaluation are driven by the choices made when addressing three major components of an evaluation strategy:

1. The method(s) and sources utilized to collect follow-up data
2. The method(s) used to isolate the effects
3. The method(s) used to convert data to monetary values

Although we should always remember the costs incurred when we make choices with these methods, the methods we ultimately use may be dictated by circumstances. These circumstances range from limited availability of the necessary data in the organization to the absence of values assigned to the organization's key measures. However, when data and sources are convenient and monetary values exist, our job as evaluators becomes much easier. The scenario below utilizing three quick evaluation methods helps to illustrate this.

- A sales training program is implemented by a major North American automobile manufacturer. The training is intended to influence sales at dealer stores in the USA and Canada. Since the training will be rolled out to more than 100 stores, management would like to evaluate several pilot offerings to determine the extent to which sales are influenced before rolling it out to all stores.

- ■ The evaluation strategy is developed as follows:
  - ■ To address the issue of isolating the effects of the training, five stores are selected to receive the training and five are selected as a control group.
  - ■ Sales data is collected from the sales records for both groups before and after the training. This requires the cooperation of only one staff financial analyst at corporate headquarters and is not labor intensive.
  - ■ The profit contribution from sales is calculated from the percentage of sales contribution factor as provided by the CFO at corporate. This is an easy calculation once the sales data are known.
  - ■ Costs are easily captured since the training is delivered by a vendor and other costs such as research, participant salaries and benefits, and travel expenses are easily estimated. It requires several hours to tabulate these costs.
  - ■ A brief questionnaire is developed and administered to determine application issues from the participants and their managers. A total of four days is dedicated to this entire process, beginning with development of the questionnaires and ending with analysis of the data.

Five to six days are dedicated to evaluating the success of this pilot offering and calculating the ROI

A TIME-SAVING WORKSHEET

Figure 13.1 provides the key questions that must be answered in order to select the most economical and credible means of collecting data, isolating the effects, and converting data to a monetary value. The methods and ideas presented in this chapter will save you time as you plan and implement evaluation processes. Share these ideas with others as well. As you consider these methods, try to track what works best for you and why. You will want to replicate the process for other evaluation initiatives.

Figure 13.1. Downloadable worksheet—Narrowing the choices.

Worksheet
Narrowing Your Choices to the Easiest and Most Practical Approach

FOLLOW-UP DATA COLLECTION	ISOLATING THE EFFECTS	CONVERTING DATA TO MONETARY VALUES
Are performance data on the trained population available from organization records?	Will logistics, economics, and ethical considerations allow the use of a control group arrangement?	Are standard values available in the organization to use for conversion?
Will a follow-up questionnaire yield the data needed? Who are the best sources and will they cooperate?	Is there more than one factor affecting the performance measures? Can a trend line analysis be used to compare before and after results?	Are cost records available for the specific measures we have identified?
Is observation on the job feasible and will it yield the data required? Will observation be too disruptive?	Are other factors likely to influence performance and do relationships exist that allow the use of forecasting methods?	Is there an internal expert who can estimate the value of the measures?
Are interviews with participants or others necessary to get the data needed and does time allow for this method?	Can participants or their supervisors estimate the program's impact? Who is in the best position to know and who will cooperate?	Is there an external expert who can estimate the value of the measures?

Figure 13.1. Worksheet—Narrowing the choices. (*continued*).

FOLLOW-UP DATA COLLECTION	ISOLATING THE EFFECTS	CONVERTING DATA TO MONETARY VALUES
Are follow-up focus groups necessary and does time allow for this method?	Can managers estimate the program's impact? Who is in the best position to know and who will cooperate?	Are there any government, industry, or research data available to estimate the value?
Does the program design lend itself to program assignments for data collection purposes?	Can subordinates report on the influence of other factors?	Are supervisors of program participants capable of estimating the value?
Is the population capable of using action planning or performance contracting to address application and data collection requirements?	Can the impact of other influencing factors be calculated or estimated?	Is senior management willing to provide an estimate of the value?
Does the program design include a follow-up session that can be used for data collection?	Are customers able to determine if skills make the difference in performance?	Does the training and development staff have the expertise to estimate the value?

CHAPTER

14

Gaining Management Support and Implementing the Process

ORGANIZATIONAL POLITICS AND DECISIONS

Political strategies in organizations take many shapes. What makes them so interesting is that people perceive them in terms of their own needs and beliefs. To some, the presence of politics in organizations points to unethical behavior. Others may view politics as necessary to keep an organization functioning in a system that does not always have clear facts or clear consequences available.

Some decisions and actions may seem political when, in reality, the decision maker(s) have facts that are unavailable to those who are judging the results of the decision. Consequently the decision seems political when, in fact, it is based on issues understood only by a few people who have a need to know. For example, when a worker is discharged for poor performance, he or she may not tell coworkers all the facts. The manager who discharges the employee is limited in what can be communicated to the work team for both legal and ethical reasons. The team members are left to draw conclusions based on incomplete facts or biased information.

For our purposes, organizational politics are defined as consequences from any action or decision by an individual or group, based on anything but the relevant facts or evidence that should influence a decision or outcome. This definition emphasizes the importance of relevant facts or evidence in the context of communicating the evaluation process and results of training programs.

Sadly, some organizations are political in their decision-making processes. That is, decisions are sometimes made based on factors other than facts that are relevant to the decision to be made. When working with such organizations, we must be even more committed to focusing the training process (including the evaluation process) on concrete and identifiable issues and needs, since strong evidence of this nature often empowers us in overcoming political agendas. Communication also becomes an important issue because the results of a training program can be closely linked to the political issues in an organization. Finally, we should never lose sight of the fact that, in many organizations, there is incredible pressure on managers and others to make decisions for political reasons. Many would rather have facts and credible evidence that allows them to make appropriate decisions.

HOW TO GET MANAGEMENT SUPPORT FOR EVALUATION OF TRAINING OUTCOMES

Senior managers are interested in any information that makes it easier for them to make decisions. The more credible the information presented, the more the likelihood that they will not have to resort to decision-making factors that may be politically charged. They will usually gravitate to data that support a logical decision. To encourage management support for training measurement (data collection), evaluation, and reporting, and to minimize political implications, we should take several actions.

Key training personnel must initiate a discussion with executive management and emphasize the necessity to collect data that demonstrate training results. This must be done in proper context. If it appears that the reason for data collection is to justify the existence of training, it will not be well received. If the purpose of data collection is communicated as implementing an organizational-results-based process, it will likely be well received.

Care also should be taken to communicate the appropriate components of results-based training, such as:

- How the evaluation of a training program can often reveal barriers to successful application/implementation of the training and how evaluation data can be used to correct these conditions
- That the lack of management support is often a primary barrier to successful implementation
- How money can be saved and results can be achieved with the ROI process
- The need for top management to support program evaluation by, as a minimum, signing memos supporting and encouraging data collection
- How asking provoking questions in executive and management meetings can help to pinpoint useful data for training objectives and evaluation

Issues in "selling" the ROI process

Some believe ROI to be a fad; others believe that it can be implemented without the necessary support that other processes require; and still others believe the ROI process can be implemented with little preparation and should not be implemented until someone important demands it. Recommended ways of communicating the desirability of the process within the organization are presented below.

When to pursue ROI. The time to pursue ROI is when you do not have to pursue it. When the current economic conditions and outlook are good, it is an excellent window of opportunity to determine results and address improvement needs. This is also a period of time when implementation is less threatening and, therefore, a more accepting environment exists.

The quick-fix mentality. The ROI process is not a quick fix. When a training department wants to implement a results-based process and establish a culture of accountability, the ROI process is an excellent choice. However, evaluating a few programs and demonstrating ROI falls far short of internalizing accountability and changing a culture.

Support of training staff. If the training staff does not see the need for ROI, it will usually fail. Trainers who have been evaluated historically by participant response sheets are not motivated to change what has been a good thing. Designers must be inspired by the challenge of a more clear linkage of training to organization goals and strategies. They must be influenced by new expectations and support to help them achieve these expectations. New roles must be provided for and understood by the entire training staff, and people must be recognized and rewarded for fulfilling these roles.

Cooperation of participants. Without the cooperation of participants, the ROI process will usually fail. Participants must be informed and inspired to provide much of the data required to complete the evaluation process. The training staff must commit themselves to finding effective ways to enlist the cooperation of participants.

Support of management. Senior management provides the funding and the ultimate expectation for results. Without the support of management, the ROI process will usually fail. This should be the group that the training staff communicates with in developing ongoing plans, gains the support of, and identifies with in terms of results.

To succeed, the ROI process requires the education of a variety of stakeholder groups, including senior managers, line managers, training managers and staff, and participants. All must understand that the overriding issue is accountability for the results of training, and that they all bear part of this accountability. They must also understand that the ROI process contributes more than financial data to help in achieving the end-in-mind accountability. They must all be informed about how the ROI process provides data on the six types of measures that are the focal point of this book.

Once the evaluation project is approved, there are several ways to keep management's approval:

Involve management early in the process. It is always helpful to educate management on the measurement process. Education is a natural by-product of involvement. One way to get early involvement of key management stakeholders is to ask them to review or set the ROI target when evaluating specific programs. This gives you the opportunity to explain the 25% hurdle rate (discussed in Chapter 2) and its conservative nature and, if necessary, to help stakeholders adjust their expectations. You should also have top management review the final values, calculations, assumptions, and methods. Again, this gives management the opportunity to clarify expectations and make necessary adjustments.

Demonstrate sensitivity to time. Good managers are almost always sensitive to the time element of resource allocation. Plan your evaluations to show concern for the time it takes others to be involved in data-collection activities. Managers often interpret research studies as time-consuming and a potential waste of resources. Clarify that you use sampling for data collection and ROI calculations with the least amount of disruption possible. Clarify that your study uses credible and practical approaches that respect the value of time while still collecting the necessary data.

Use reliable sources. When converting data to values, use estimates from the most reliable and credible sources. You must be viewed as the messenger and as an objective research analyst. The results you present must be viewed as credible when placed alongside other data that management views on a daily basis. For this reason, it is important to tap these same sources when appropriate and possible.

Use a conservative approach with data. Take a conservative approach to developing the monetary values that are presented as the benefits from the training. When presented with alternative values during data analysis, always choose the most conservative value. When using estimates of improvements, be sure to make adjustments for potential error. Also, be sure to capture all costs, and use the conservative approach when estimating costs.

Show how programs are producing needed results. When addressing results, there are two areas of concern to managers. Did the training help solve a major problem (or connect to an opportunity) and, in reporting results, did we account for the influence of other factors? Your study should do both, and these issues should be highlighted as you plan and report the results. When you demonstrate that a program solves a major problem, you communicate that the program is based on specific needs. As you communicate how you isolate the effects, you demonstrate that you understand that other factors are also influencing outcomes. Of course, when a program does not achieve results, your communication should focus on the need for improvement in the training process.

Taking the above actions will not guarantee an evaluation process free of political decisions or implications. However, presenting facts and evidence in these ways can help stakeholders to put political agendas (e.g., funding for competing programs) aside in favor of making decisions that will benefit the organization. When the credibility of a process allows it to withstand the close scrutiny of numerous stakeholders, the needs of all are served.

AVOIDING THE MISUSE OF MEASUREMENT

Measurement should be used to address the targeting of goals and objectives and the reporting of progress and results. As such, measurement is a useful tool to help organizations and individuals know where they are going and to determine when they have arrived. But when measurement is used as a means for political gain, it can be a destructive force in the organization. The following should be avoided.

Avoid Measurement to Advance Careers. On occasion, measurement can be viewed by a sponsor stakeholder as the edge needed to advance a career. When this political motive is the reason measurement is introduced into the organization, the process usually fails. The sponsor usually exhibits destructive behavior. For example, the sponsor may try to force the evaluation process without paying attention to resistance from other stakeholders. Resistance often occurs during any change because certain groups of stakeholders either do not understand, or they disagree with the purpose of the change sponsors. Often, substantial resources can be squandered before the process is stopped or fails of its own weight. To prevent the misuse of measurement, the process should be communicated as a change initiative that requires the same nurturing and sponsorship as would any major change. This begins with education and preparation of appropriate communication plans for timely implementation. Communication should address the purpose of measurement as a means to improve quality and results. Sponsor and stakeholder behavior should be consistent with this purpose. All stakeholders should understand that the measurement process is not an invasive or finger-pointing exercise, but a discovery process to achieve improved results. When a sponsor insists on driving the process in an arbitrary, autocratic, or unreasonable manner, the process should be aborted if it cannot be salvaged and redirected.

Avoid Measurement to Discover Fault. The focus of the measurement process should be on outcomes (the desired ones or others) and the processes that contributed to the outcomes. When data are collected, analyzed and reported, it is imperative that the desired outcomes (usually expressed in terms of improvement goals) be in the forefront. Occasionally, sponsors or stakeholders may exhibit misguided behavior (or sometimes unethical behavior) by communicating the process as a fault-finding mission. Others then draw conclusions that negative implications or punishment will be forthcoming as they try to guess the motives of the sponsor. This approach contributes to mistrust in the organization and often results in failure for the measurement process.

Continuous education of potential sponsors, including speaking directly to sponsorship issues, can increase the probability of a smooth installation of measurement processes. This education process should be repeated in the form of a briefing each time a measurement project is initiated. It is also advisable to establish policies around these issues and get stakeholder buy-in of the policies. All stakeholders should be included and the briefing should be customized to meet the needs of the situation.

IMPLEMENTATION

As with any process, effective implementation is the key to success. Implementation occurs when the new process is effectively integrated into the routine of the organization. Without effective implementation, even the best process will fail. There must be clear-cut steps for a comprehensive implementation process that will overcome resistance. This works best when the process involves accountability.

Even the best-designed process or technique is worthless without this integration. Often, there is resistance to the ROI process, both from the client and those involved in the training process. Some of this resistance is based on fear and misunderstanding. Some is based on actual barriers. Although the ROI process presented in this book is methodical, it can fail if it is not integrated properly and fully accepted and supported by those who must make it work in the organization.

Overcoming resistance

With any new process or change, there is resistance. Resistance may be especially great when implementing a process as complex as ROI that focuses on accountability. There are three key reasons why we should address ways to overcome resistance.

1. Universal resistance to change. Resistance is always present when change is being implemented. Sometimes there are good reasons for it, and often there are not. The important thing is to sort out both types and try to dispel any myths. When legitimate barriers are the basis for resistance, trying to minimize or remove them altogether is the challenge.

Sponsorship at the highest level is useful in overcoming this type of resistance.

2. Scarce resources (e.g., staffing and budget) always play a role in resistance. The approaches described in Chapter 13 will be helpful here.

3. Leadership support at all levels (individual, group, cross-functional team) is focused on many priorities. To overcome the natural resistance associated with conflicting priorities, we must provide leaders with clear reasons why ROI data collection and measurement deserves their time and support. Here, we can focus on the value that measurement brings to the allocation of scarce resources. By providing guidance toward the organization's objectives, measures actually help prevent the waste of precious time, resources, and misguided outcomes, and therefore bring efficiency and effectiveness to the organization.

Key implementation steps

1. Determine/establish responsibilities. Specific roles and responsibilities of major stakeholders must be clearly defined so the measurement and evaluation process is implemented effectively and is successful. Responsibilities are shared not only within the training department but with other, closely related groups, including the business units, participant managers, and participants.

The policy examples below are taken from the training evaluation manual of a major pharmaceutical firm.

Section 1.1 Overall Responsibility (Downloadable Form)

Measuring the effectiveness of training programs is a shared responsibility of several important groups. The training department, through the direction of sales operations management, has the overall responsibility to:

■ Manage the evaluation process from the needs assessment to communicating the results of evaluation.

- Ensure that an appropriate needs assessment is conducted before proceeding with a new training program.
- Identify application objectives (Level 3) and business impact objectives (Level 4) for each new training program.
- Set targets for evaluation for all programs at each level.
- Develop an evaluation strategy for each training program.
- Design training programs with evaluation procedures in place to measure results.
- Provide participants and managers information about the evaluation process and expected results from each training program.
- Coordinate the data-collection process.
- Coordinate or conduct data analysis.
- Interpret results and review evaluation reports.
- Communicate evaluation results to selected target audiences.

Section 1.3 Instructional Designer Responsibilities (Downloadable Form)

The Instructional Designer should:

- Ensure that application objectives and business impact objectives are developed for each new program before the design process begins.
- Develop content that is relevant to job settings and linked to business unit issues.
- Design exercises and activities that link to the business environment, needs, and issues.
- When possible, develop evaluation tools within the training program for self-assessment and for reporting results.
- Communicate the results-based focus to vendors who design and develop training programs.
- Revise and re-design programs based on results of previous evaluations.

Section 1.6 Participants' Responsibilities (Downloadable Form)
Participants must:

- Participate fully in training programs to learn as much as possible.

- Explore ways in which learning can be applied on the job.

- Partner with manager to choose training programs that can best improve business performance.

- Enter into the training program with an open mind and be willing to learn new concepts and develop new skills.

- Take responsibility for success of the application of the training program.

- When requested, provide information and feedback on success of the training program and the barriers to implementation.

- Partner with manager to identify and remove barriers to the application of training programs.

- Have the determination to achieve success with the training program in the face of many obstacles to application and improvement.

2. Educate managers and others on the ROI process. There are several key target audiences for education about evaluation. As addressed earlier, executive management is a key target for education and should receive at least a briefing on the process. Middle-level managers are a key influence on training and should be targeted for education, perhaps as much as four to eight hours. In some organizations, a primary target for ROI education may be first-level managers, and in others, the target may begin with second-level managers. Three important questions help determine the proper audience:

QUESTIONS TO HELP DETERMINE THE PROPER AUDIENCE

- Which group has the most direct influence on the training function?

- Which group is causing serious problems with lack of management support?

- Which group has the need to understand the ROI process so they can influence training transfer?

In addition to the above stakeholders targeted for education, the training staff is a primary target. The knowledge of and skill in evaluation that is required is determined by the role each staff member performs in the training and evaluation processes.

3. Develop transition/implementation plan. While it takes only a few months to design and implement an impact study, the implementation of the results-based approach will not occur quickly. It usually evolves over time and will likely take several years to be fully operational. The timing of the implementation is very critical because changes in philosophy and approaches are needed, along with a redirection of resources. To keep the implementation on target and the training staff in focus on this important transition, a leading telecommunications firm set the targets shown in Figure 14.1.

The above targets may seem ambitious but are essential to make the transition to a results-based approach within a reasonable time frame.

4. Prepare/revise evaluation policy and guidelines. Because the results-based process represents a change for the training organization, it is important to review the evaluation policies and guidelines and install new procedures compatible with the goals of the ROI evaluation process. Figures 14.2, 14.3, and 14.4 include portions of the evaluation guidelines for a government agency.

5. Set targets for evaluation levels. To ensure that there is a comprehensive system of evaluation while considering the resources available for evaluation, specific targets should be set for each level of evaluation. This process requires that a certain number of programs be evaluated at each level. Figure 14.5 shows an example of how one company set these targets.

These targets provide guidance for the training department to focus on a complete evaluation system. For evaluations planned at

Figure 14.1. (Downloadable form.) Results-based transition plan.

1. By the end of 2002, all training programs will have a data-collection plan in place. The plan will include the following items:
   - Objectives
   - Data-Collection Methods
   - Timing
   - Responsibilities
2. By the end of 2003, the following evaluation targets will be established:
   - 100% of the training products will be evaluated at Level 1.
   - 50% of the training products will be evaluated at Level 2.
   - 30% of the training products will be evaluated at Level 3.
   - 20% of the training products will be evaluated at Level 4.
   - 10% of the training products will be evaluated at Level 5.
3. By the end of 2003, all new training programs will be designed to have the capability to be taken to a Level-5 evaluation.
4. As new programs are designed or purchased, evaluation at levels 3 and 4 will include a strategy for isolating the effects of the training program. Evaluation at Level 5 will include strategies for converting the data to monetary values and tabulating the cost of the program.
5. By the end of 2002, a results-based cross-functional needs assessment will be conducted for 100% of new training programs sponsored by the Corporate University. This will require that a diagnosis/needs assessment be focused on Level 3 and 4 data and from this data, Level-3 and -4 objectives for the training programs will be developed.

Level 5, a Level 4, 3, 2, and 1 evaluation must also be planned to ensure that a chain of impact has occurred that indicates that participants learned the material, applied it on the job, and obtained the desired results.

6. Identifying a champion. As a first step, one or more individual(s) should be designated as the internal leader or champion for measurement and evaluation. As in most change efforts, someone must take the responsibility for ensuring that the process is imple-

Figure 14.2. (Downloadable form.) New and proposed programs.

Guideline for New and Proposed Training Programs

The results-based approach will have a dramatic effect on new or proposed training programs. Because training programs must be linked to business needs, a thorough needs assessment will be required. A precise business need must be identified that will be enhanced, influenced, or corrected by the training program. In addition, specific behavioral changes needed in the future must be identified. This requirement moves needs assessment beyond the traditional assessment of skills, knowledge, or attitudes to the assessment of on-the-job performance and business problems and opportunities. With this approach, a new or proposed training program will not be pursued unless it directly addresses a legitimate business need and changing behavior linked to the business need.

A second element in this implication is that new or proposed training programs will have Level-3 and -4 objectives developed for them. Specific outcomes will be developed to link to desired behavioral change and business impact. Specific and measurable objectives provide the driving force for training program design, development, delivery, and evaluation. More importantly, these objectives provide focus for participants to enhance job performance and improve business results. After these are developed, the new or proposed training program can proceed in design and development stages.

Figure 14.3. (Downloadable form.) Existing programs.

Guideline for Existing Training Programs

The results-based approach will affect existing training programs in several ways. First, a few training programs will be selected for increased evaluation activity at Levels 3, 4, and 5. Based on the results from evaluation data, the training program could be enhanced, modified, changed, or discontinued. Second, facilitators for ongoing programs will be asked to relate learning activities more closely to output measures. In some training programs, specific action plans will be developed to link programs to output variables. Third, when a training program is being revised or redesigned, the needs assessment will be revisited and Level-3 and -4 objectives will be established.

The results-based philosophy will be integrated into the existing programs on a gradual basis, and ultimately all training programs will have the results-based approach fully in place.

---

mented successfully. This leader is usually the one who understands the process best and sees the potential contribution of the process. More importantly, this leader is willing to teach others.

The measurement and evaluation leader is usually a member of the training staff who has this responsibility full time (in larger organizations) or part-time (in smaller organizations). The typical job title for a full-time leader is manager of measurement and evaluation. Some organizations assign this responsibility to a team and empower it to lead the effort.

7. Assessing the climate. A final element in planning the implementation is to assess the current climate for achieving results. Two groups are targeted for this effort: the internal training or corporate university staff and the management group.

Figure 14.4. (Downloadable form.) New skills and knowledge.

Guidelines for New Skills and Training

This results-based approach requires a paradigm shift for most individuals involved in the training process. Consequently, new skills and knowledge must be acquired.

Training team members must understand the results-based approach so that they will know how to apply and support it.

The directors of training need to understand the results-based approach and their roles in making it successful.

Trainers must understand the approach so that they can utilize results-based activities as they facilitate programs.

Training vendors and suppliers need to understand the approach so they can develop training programs linked to business results.

Managers of participants must understand their role in making training successful, as they reinforce and support learning on the job with a new frame of reference.

Senior executives must have enough exposure to the results-based approach to understand how training programs will contribute to overall organizational goals.

Figure 14.5. Evaluation targets.

LEVEL OF EVALUATION	PERCENT OF PROGRAMS EVALUATED AT THIS LEVEL
Level 1–Reaction and Planned Action	100%
Level 2–Learning	50%
Level 3–Job Applications	30%
Level 4–Business Impact	20%
Level 5–Return on Investment	10%

Because of the importance of support from the management group, it is helpful to assess manager perception of the effectiveness of the training function. An instrument designed specifically for management input is presented in Figure 14.6, along with an interpretation of scores (Figure 14.7). An assessment process provides an excellent opportunity to discuss current issues and concerns. A gap analysis can reveal specific areas for improvement and goal setting.

Then the organization can plan for significant changes, pinpointing particular issues that need improvement and support as measurement and evaluation is enhanced. In addition, some organizations take annual assessments to measure progress.

Figure 14.6.

Training and Development Programs Assessment:
A Survey for Managers

Instructions. For each of the following statements, please circle the response that best describes the training and development function in your organization. If none of the answers describe the situation, select the one that best fits. Please be candid. Select the most accurate response.

1. The direction of the training and development function in your organization:
   a) Shifts with requests, problems, and changes as they occur.
   b) Is determined by Human Resources and adjusted as needed.
   c) Is based on a mission and a strategic plan for the function.

2. The primary mode of operation of the training and development function is:
   a) To respond to requests by managers and other employees to deliver training programs and services.
   b) To help management react to crisis situations and reach solutions through training programs and services.
   c) To implement many training programs in collaboration with management to prevent problems and crisis situations.

3. The goals of the training and development function are:
   a) Set by the training staff based on perceived demand for programs.
   b) Developed consistent with human resources plans and goals.
   c) Developed to integrate with operating goals and strategic plans of the organization.

4. Most new programs are initiated:
   a) By request of top management.
   b) When a program appears to be successful in another organization.
   c) After a needs analysis has indicated that the program is needed.

5. When a major organizational change is made:
   a) We decide only which presentations are needed, not which skills are needed.
   b) We occasionally assess what new skills and knowledge are needed.
   c) We systematically evaluate what skills and knowledge are needed.

6. To define training plans:
   a) Management is asked to choose training from a list of canned, existing courses.
   b) Employees are asked about their training needs.
   c) Training needs are systematically derived from a thorough analysis of performance problems.

7. When determining the timing of training and the target audiences:
   a) We have lengthy, nonspecific training courses for large audiences.
   b) We tie specific training needs to specific individuals and groups.
   c) We deliver training almost immediately before the skills are to be used, and it is given only to those people who need it.

Figure 14.6. (*continued*)

8. The responsibility for results from training:
   a) Rests primarily with the training staff to ensure that the programs are successful.
   b) Is a responsibility of the training staff and line managers, who jointly ensure that results are obtained.
   c) Is a shared responsibility of the training staff, participants, and managers all working together to ensure success.

9. Systematic, objective evaluation, designed to ensure that trainees are performing appropriately on the job:
   a) Is never accomplished. The only evaluations are during the program and they focus on how much the participants enjoyed the program.
   b) Is occasionally accomplished. Participants are asked if the training was effective on the job.
   c) Is frequently and systematically pursued. Performance is evaluated after training is completed.

10. New programs are developed:
    a) Internally, using a staff of instructional designers and specialists.
    b) By vendors. We usually purchase programs modified to meet the organization's needs.
    c) In the most economical and practical way to meet deadlines and cost objectives, using internal staff and vendors.

11. Costs for training and OD are accumulated:
    a) On a total aggregate basis only.
    b) On a program-by-program basis.
    c) By specific process components such as development and delivery, in addition to a specific program.

12. Management involvement in the training process is:
    a) Very low with only occasional input.
    b) Moderate, usually by request, or on an as-needed basis.
    c) Deliberately planned for all major training activities, to ensure a partnership arrangement.

13. To ensure that training is transferred into performance on the job, we:
    a) Encourage participants to apply what they have learned and report results.
    b) Ask managers to support and reinforce training and report results.
    c) Utilize a variety of training transfer strategies appropriate for each situation.

14. The training staff's interaction with line management is:
    a) Rare. We almost never discuss issues with them.
    b) Occasional; during activities such as needs analysis or program coordination.
    c) Regular; to build relationships as well as to develop and deliver programs.

15. Training and development's role in major change efforts is:
    a) To conduct training to support the project, as required.
    b) To provide administrative support for the program, including training.
    c) To initiate the program, coordinate the overall effort, and measures its progress, in addition to providing training.

Figure 14.6. (*continued*)

16. Most managers view the training and development function as:
    a) A questionable function that wastes too much employee time.
    b) A necessary function that probably cannot be eliminated.
    c) An important resource that can be used to improve the organization.

17. Training and development programs are:
    a) Activity-oriented (e.g., All supervisors will attend the "Performance Appraisal Workshop.")
    b) Individual-results-based (e.g., The participant will reduce his or her error rate by at least 20%.)
    c) Organizational-results-based (e.g., The cost of quality will decrease by 25%.)

18. The investment in training and development is measured primarily by:
    a) Subjective opinions.
    b) Observations by management and reactions from participants.
    c) Dollar return through improved productivity, cost savings, or better quality.

19. The training and development effort consists of:
    a) Usually one-shot, seminar-type approaches.
    b) A full array of programs to meet individual needs.
    c) A variety of training and development programs implemented to bring about change in the organization.

20. New training and development programs, without some formula method of evaluation, are implemented at my organization.
    a) Regularly
    b) Seldom
    c) Never

21. The results of training programs are communicated:
    a) When requested, to those who have a need to know.
    b) Occasionally, to members of management only.
    c) Routinely, to a variety of selected target audiences.

22. Management involvement in training evaluation:
    a) Is minor, with no specific responsibilities and few requests.
    b) Consists of informal responsibilities for evaluation, with some requests for formal training.
    c) Is very specific. All managers have some responsibilities in evaluation.

23. During a business decline at my organization, the training function will:
    a) Be the first to have its staff reduced.
    b) Be retained at the same staffing level.
    c) Go untouched in staff reductions and possibly be beefed up.

24. Budgeting for training and development is based on:
    a) Last year's budget.
    b) Whatever the training department can "sell."
    c) A zero-based system.

Figure 14.6. (*continued*)

25. The principal group that must justify training and development expenditures is:
    a)   The Training and Development department.
    b)   The human resources or administrative function.
    c)   Line management.

26. Over the last two years, the training and development budget as a percent of operating expenses has:
    a)   Decreased
    b)   Remained stable
    c)   Increased

27. Top management's involvement in the implementation of training and development programs:
    a)   Is limited to sending invitations, extending congratulations, passing out certificates, etc.
    b)   Includes monitoring progress, opening/closing speeches, presentation on the outlook of the organization, etc.
    c)   Includes program participation to see what's covered, conducting major segments of the program, requiring key executives to be involved, etc.

28. Line management involvement in conducting training and development programs is:
    a)   Very minor; only training specialists conduct programs.
    b)   Limited to a few specialists conducting programs in their areas of expertise.
    c)   Significant. On the average, over half of the programs are conducted by key line managers.

29. When an employee completes a training program and returns to the job, his or her manager is likely to:
    a)   Make no reference to the program.
    b)   Ask questions about the program and encourage the use of the material.
    c)   Require use of the program material and give positive rewards when the material is used successfully.

30. When an employee attends an outside program/seminar, upon return, he or she is required to:
    a)   Do nothing.
    b)   Submit a report summarizing the program.
    c)   Evaluate the program, outline plans for implementing the material covered, and estimate the value of the program.

Figure 14.7.

Interpreting the Training and Development Programs Assessment

Score the assessment instrument as follows. Allow

  1 point for each (a) response.

  3 points for each (b) response.

  5 points for each (c) response.

The total will be between 30 and 150 points.

The interpretation of scoring is provided below. The explanation is based on the input from dozens of organizations and hundreds of managers.

Score Range	Analysis of Score
120–150	**Outstanding Environment** for achieving results with training and development. Great management support. A truly successful example of results-based training and development.
90–119	**Above Average** in achieving results with training and development. Good management support. A solid and methodical approach to results-based training and development.
60–89	**Needs Improvement** to achieve desired results with training and development. Management support is ineffective. Training and development programs do not usually focus on results.
0–59	**Serious Problems** with the success and status of training and development. Management support is non-existent. Training and development programs are not producing results.

Index

Note: **Boldface** numbers indicate illustrations.

ABOUT THE AUTHORS

Jack J. Phillips, Ph.D., is one of today's leading authorities on training and performance measurement. He is with the Jack Phillips Center for Research, a division of Franklin Covey. He is a veteran author and expert on human resource management and provides consulting services for Fortune 500 companies in the United States and organizations in 25 countries. He is author or editor of more than 32 books and 100 articles.

Ron Drew Stone is the practice leader for the Jack Phillips Center for Research. He directs the consulting practice in measurement and accountability. He consults regularly with Fortune 500 companies and a wide range of international clients.